Jews in American Agriculture: An Annotated Bibliography

Jews in American Agriculture: An Annotated Bibliography

IRWIN WEINTRAUB

Foreword by RONALD BECKER

McFarland & Company, Inc., Publishers
Jefferson, North Carolina, and London

LIBRARY OF CONGRESS CATALOGUING-IN-PUBLICATION DATA

Jews in American agriculture : an annotated bibliography /
Irwin Weintraub ; foreword by Ronald Becker.
p. cm.
Includes bibliographical references and index.

ISBN-13: 978-0-7864-2947-9
(softcover : 50# alkaline paper) ∞

1. Jewish farmers — United States — Bibliography.
2. Jews — United States — Bibliography. I. Weintraub, Irwin.
Z5075.U5J49 2007
[HD1525]
016.63092'2924073 — dc22 2006101088

British Library cataloguing data are available

Cover photograph: Jewish family harvesting bean crop, Seabrook Farm, Bridgeton, NJ, 1942 (Library of Congress)

Manufactured in the United States of America

McFarland & Company, Inc., Publishers
Box 611, Jefferson, North Carolina 28640
www.mcfarlandpub.com

Acknowledgments

I would like to thank the libraries, research centers, and archives that have extensive resources on Jewish life and culture in the United States, Israel and the world. Without their assistance, many of the sources cited here would not have come to my attention.

I began my research at Brooklyn College Library, which has a wide range of books and periodicals on all aspects of Jewish life and culture. I am indebted to Sherry Warman and Theresa Ferrara from Brooklyn College Library's Interlibrary Loan office. They were a great help in finding many of the needed resources in libraries and archives around the country. Sherry pointed out many resources that I missed and helped me acquire them.

Whenever I have a research topic to pursue, I visit the New York Public Library (NYPL) and never cease to be amazed at the scope of that library's resources. The NYPL is truly a powerhouse of information for any researcher. I found many of the works cited here at the Dorot Jewish Division. Librarians Miriam Gloger and Eleanor Yadin got caught up in the excitement of this topic with me and saw to it that I did not miss anything that was in the NYPL collection related even in the smallest way to Jews in American agriculture. Miriam and Eleanor are great librarians who know their collections well and serve their patrons with the skill and enthusiasm that has made NYPL a library of renown.

The Center for Jewish History in New York, which contains in one location the American Jewish Historical Society, Yivo Institute for Jewish Resarch, Leo Baeck Institute for the Study of German-Speaking Jewry, American Sephardi Federation and Yeshiva University Museum, provided many archival documents that were not available elsewhere. I recommend that all who are interested in Jewish life and history visit the Center. See the wonderful exhibits, do research in the reading room, and browse the Center's bookstore.

Thanks to Jeanne Abrams at the Rocky Mountain Jewish Historical Society for the latest information about the Cotopaxi, Colorado, agricultural settlement, and to Andrew Muchin, director of the Wisconsin Small Jewish Communities History Project at the Wisconsin Society for Jewish Learning in Milwaukee for the articles about Arpin, Wisconsin.

A special thank you to Ronald Becker, Head of Special Collections and University Archives at Rutgers University libraries for writing the foreword. Ron grew up on a poultry farm in Vineland, New Jersey, and has first hand knowledge of life in a Jewish agricultural colony. His foreword will open the door for readers who want to enter the world of Jewish agriculture and learn more about its legacy.

I would like to thank Lawrence Albrecht, evening and weekend manager of the Brooklyn College Library Café, for compiling the title and subject indexes. Larry is a computer whiz who welcomes the challenges of technology and solves them with ease.

Contents

...then I will give your rains in their season, and the land shall yield her produce, and the trees of the land shall yield their fruit. And your threshing shall reach unto the vintage, and the vintage shall reach until the sowing time; and you shall eat your bread until you have enough, and dwell safely in your land. And I will give peace in the land, and you shall lie down and none shall make you afraid.... (Leviticus 26:4–6)

✡

For the Lord thy God bringeth thee into a good land, a land of brooks of water, of fountains and depths, springing forth in valleys and hills; a land of wheat and barley, and vines and fig trees and pomegranates; a land of olive trees and honey; a land wherein thou shalt eat bread without scarceness, thou shalt not lack any thing in it; a land whose stones are iron, and out of whose hills thou mayest dig brass. And thou shalt eat and be satisfied, and bless the Lord thy God for the good land which he hath given thee. (Deuteronomy 8:7–10)

✡

...I will give the rain of your land in its season, the former rain and the latter rain, that thou mayest gather in thy corn, and thy wine and thine oil. And I will give grass in thy fields for thy cattle, and thou shalt eat and be satisfied. (Deuteronomy 11:14–15)

Foreword

by Ronald Becker

The contributions of Jewish Americans to agriculture and rural society are many and varied and are well documented in a myriad of publications, some of which have been widely disseminated. In addition, there is ample archival documentation in the papers of the Jewish Agricultural Society, the Baron de Hirsch Fund, and other collections. These contributions have been sustained throughout American history and in all regions of the country, with the largest impact beginning in the 1880s and lasting nearly a century. Jewish farmers and agricultural scientists made lasting contributions to their professions, to their communities (both Jewish and secular), and to the nation far beyond their actual number. In addition, farm life provided the close community ties and resources for Jewish pioneers from which they have raised new generations of highly productive members of American society. The offspring of Jewish farming families have made a mark in every aspect of cultural, social, economic, educational, and civic life in the United States. Examples include U.S. Senator Arlen Specter who is descended from a Jewish farming family in Kansas. Also from Kansas, Daniel Glickman served as U.S. Secretary of Agriculture, prior to which he represented his district in the U.S. House of Representatives for eighteen years.

This volume demonstrates that Jews entered agricultural life from many different backgrounds and with a variety of motivations. Some were farmers in Europe and elsewhere and came to America with the desire to farm here. Other immigrants with little or no farming experience came to America nearly penniless and were attracted by the opportunities in agrarian communities. Others began working in the factories and stores of urban America, and were lured to the farm to escape the stifling life of the crowded cities, and trade it in for the fresh air of the countryside and the opportunity to own property. Although the majority of the farms were on the East Coast, especially in New Jersey, New York, and Pennsylvania, there were Jewish agrarian settlements in every region of the country.

My own history began in a small, crowded walk-up apartment in the South Bronx from which my family relocated to a poultry farm in Vineland, New Jersey, where I grew up. In the city, my immigrant parents toiled in the garment industry, my mother as a hand sewer paid by the piece, and my father as a presser, working under a huge hot iron for many continuous hours day after day. The work was difficult, the pay was poor, the English language was bewildering, and the commute was stifling (my father traveled the subways from the Bronx to Brooklyn). We were six people (my parents, myself, my mother's parents, and my mother's brother) during the week but expanded to eleven on the Sabbath when my mother's sisters and their families came to spend the night and following day. All crammed into a two-bedroom apartment! My father often complained that he didn't survive the Holocaust for this type of life.

One day in the summer of 1951, my parents embarked on a 3½ hour bus ride to visit an old friend of my mother in Vineland. She had married a fellow immigrant and settled on a poultry farm. The weather was fine and so was the company, the fresh air, and the aroma of the barbecue. Without knowing any of the details of farm work, my father determined then and there that he would be joining these people as soon as he could. The next day, instead of commuting to his job in Brooklyn, he got off the subway near the headquarters of the Jewish Agricultural Society on Park Avenue in New York, and arranged to get a farm in Vineland. Despite the extended family's shock and bewilderment, we did indeed move there within a very short time.

My parents were greeted by a dilapidated bungalow, mosquitoes, floods, backbreaking work, and even longer working hours than they endured in the factory. The chickens didn't take off for the Sabbath and diseases spread through the flocks regularly. Yet despite all this, my parents made a wonderful life from the farm, raised two children who both became professionals and active members of their secular and religious communities, and died knowing that they had made solid contributions to their adopted country.

Even though that particular way of life in Vineland is essentially gone, its impact is all around us. Hundreds of other families survived the first generation in America on the farm and set the stage for the generations that have followed, and will follow.

My own upbringing is only one of many diverse experiences on the American farm. The reasons for Jews to enter agriculture were many. Many residents of the small shtetls of Europe were engaged in subsistence farming in one way or another. Even those involved in commerce or peddling often had a few chickens, grew fruits and vegetables, and, if they were lucky, had a cow or even a horse. They were used to hard work and simple living and saw opportunities for farming in America as a continuation of that aspect of their former lives. Others were zealous Zionists who believed that farming in America would be the first step towards eventual emigration to the land of Israel. Still others were attracted to the romance of farming and rural life much the same as the non-Jewish Americans who populated the agricultural communities. Small privately-owned and privately-run farms were generally preferred to collectives and utopian communities, although there are examples of these as well. However, Jewish farmers tended not to be loners. They usually sought existing rural hamlets, or, where none existed, created their own. Once settled, they established schools, synagogues, community and civic centers, social associations, and organizations providing technical support for farming activities.

Both immigrant and second-generation settlers made their mark in agricultural and scientific research, business, and scholarship. Several Jewish farmers went on to become county agricultural agents. Small schools were established, expanded, and later superseded by or incorporated into university programs. For example, in southern New Jersey, the Woodbine Agricultural School served a small but burgeoning community of growers and chicken farmers. Soon after, the National Farm School was founded in Doylestown, Pennsylvania. Renowned soil chemist and bacteriologist Jacob Lipman immigrated to the U.S. from eastern Europe and began his education at the Woodbine school, and applied his knowledge and training to the surrounding farms. He later directed the regional agricultural experiment station, and ultimately served as dean of the Rutgers University College of Agriculture. His siblings and later generations also made sustained contributions to American agriculture and education. Another pioneer in American agriculture was Arthur Goldhaft, who established the Vineland Poultry Laboratory where he developed vaccines used internationally to eradicate what had been devastating poultry diseases. Jacob Rubinoff expanded his small farm supply operation in southern New Jersey to establish his own advanced feed formula. Most renowned was Selman Waksman, who received his degree in agricultural science from Rutgers and then was invited back by Dean Lipman to serve as professor of soil bacteriology. He and his students isolated numerous new antibiotics — most notably streptomycin and neomycin, which have fought infectious diseases for decades. Rutgers later named its Institute for Microbiology in his honor.

The children of pioneering Jewish poultry farmers succeeded in professional life as well. Raised in the small Jewish farming community of Alliance, New Jersey, Gilbert Seldes served as a correspondent during World War I and went on to become dean of the School of Communication Arts at the University of Pennsylvania. Also from Alliance, Ben Golder became the first Jewish congressman to be elected from Philadelphia. Ben Leuchter became the chief editor of the *Vineland Times Journal*, the largest daily newspaper in the region. Polish immigrant Miles Lerman succeeded in business and became the driving force behind the establishment of the U.S. Holocaust Museum. Jewish writers and poets recounted their early years in small Jewish farming communities in newspapers and periodicals.

Other first and later generations of Jewish residents in rural American communities devoted their lives to serving the farming communities. They worked in the feed mills, the farmers' co-ops, in post offices and other service agencies, and in retail. They also started kosher food establishments, became teachers, entered various professions, and became religious and community leaders. Some entrepreneurs in the Catskill Mountain region of New York and other well-situated places served a larger community by establishing resort hotels so those living in the hot cities could enjoy comfortable summer vacations. No complete history of my hometown or many others could be written without paying particular attention to the contributions of Jewish settlers.

Yet despite the many contributions contained in primary and secondary sources, scholars and laypersons alike generally tend to ignore or give only brief mention to them whether writing about the history of agriculture or the history of the Jewish experience in the United States. In *Jews in American Agriculture: An Annotated Bibliography*, Irwin

Weintraub has produced a valuable volume that provides the opportunity for researchers to gain access to sources and thereby expand our knowledge of Jews in agriculture throughout American history. As former head of the Stephen and Lucy Chang Science Library at Rutgers, which specializes in all aspects of agricultural sciences, and for the scholarly work he has produced throughout his career as a librarian, Weintraub typifies the very contributions of Jews to the practice and study of the agriculture alluded to in this foreword.

Ronald Becker is Head, Special Collections and University Archives, Rutgers University.

Preface

Farmers who work the land and provide food, fiber, and raw materials are unsung heroes who make enormous contributions to the societies in which they live. Those yeomen are the major players in an interrelated system of producers, processors, distributors, and merchants who supply consumers with products they need for survival.

Agriculture has always been a significant part of religious and secular aspects of Jewish life. The food and harvest festivals, kosher dietary laws, and the commandments and tenets regarding stewardship of land and humane treatment of animals throughout scriptures involve agricultural decision-making year round.

I grew up in the urban environment of Brooklyn, New York, far removed from farms and farmers. During summers spent in the country, I visited farmers and observed the many roles they perform each day to bring food to American dinner tables. I was fascinated by their relationships with the land and animals and the practical knowledge and skills they must possess to successfully manage a farm business.

Following graduation from high school, I decided to major in agriculture and be a part of this noble profession. I acquired master's degrees in agriculture and rural sociology and spent two years in the Peace Corps as an agricultural agent in northwestern Kenya's Turkana district. When I returned from Kenya, I enrolled in a library science program and received a master's degree in Library Science and years later a Ph.D. in Library and Information Studies. My career includes over 25 years as an agriculture librarian in land grant colleges in the United States and several years in Israel. As an agriculture information specialist, I have witnessed the enormous exchange of information that takes place daily among all segments of the industry in the United States and abroad.

Among that vast body of literature are historical and sociological studies of the ethnic groups engaged in farming. Much has been written about Scandinavian, German, Mennonite, Amish and African American farmers. By contrast, historical works on American Jews tend to minimize or ignore the agricultural sector as a phenomenon of little importance. Some historians focus on the failures and conclude that Jews are not suited for life on the farm. Those authors don't seem to be aware of the successful Jewish farmers around the country, the Jewish teachers and professors in agricultural schools and colleges, and the large number of Jews involved in the processing, handling, and distribution of farm products.

My goal in compiling this bibliography is to explore the public record documenting the Jewish contribution to the history and development of American agriculture. This is not an obscure record hidden away in somebody's attic. The literature depicts a long and ongoing legacy of Jews as landowners, independent farmers, ranchers, cowboys, teachers, extension agents, scientists, consultants, and even a Jewish secretary of agriculture.

The story of Jewish farmers and their lives can fill many bibliographies and historical treatises. There are articles and publications in local newspapers, specialty magazines, and archives that were not cited here. But this bibliography will serve as a framework to educate, enlighten, and raise awareness about a forgotten aspect of Jewish life in America. I look forward to hearing from readers who have stories to tell about Jewish farmers that they know and the effect those individuals had on their lives.

Introduction

Agriculture is the foundation of every existing civilization and has been an inseparable part of Jewish life and history since ancient times. When Jews were wandering in the desert as a free people, it was necessary that they employ the best agricultural practices for sustenance during the long journey. God's promise to deliver the Jewish people to "a land flowing with milk and honey" is a metaphor that emphasizes the importance of agriculture to the Jewish soul. There are passages throughout the Bible regarding humane conservation and stewardship practices that keep the land viable and productive for future generations. The holidays of Shavuoth and Succoth celebrate abundant harvests and agricultural renewal. Jewish Arbor Day, also known as "New Year of the Trees," celebrated on the 15th day of the Jewish month of Shevat, is commemorated by eating fruits and planting trees. This holiday is a time to reflect on the environment and our responsibility to protect the fragile ecosystem. In the Talmud, large portions of the *Mishna* relate to agricultural practices. The extraordinary transformation of Israel from a semi-arid wasteland to an agricultural horn of plenty is a shining example of the Jewish relationship to the land and its potential to produce a bountiful harvest for over six million of its inhabitants and a flourishing food export industry.

This love of the land and savvy agricultural know-how followed the Jewish people to the countries of the diaspora. Jews have been farming in America since the first settler arrived 350 years ago. One of the most active periods began in the 19th century. In the 1830s, a wave of immigrants from Germany and Poland arrived in cities and rural areas in the South, Midwest, and West. Many of those settlers became large-scale farmers and ranchers, acquired extensive holdings, and established a strong agricultural presence in their communities. When oppressive governments in Eastern Europe decreed that Jews were forbidden to own land, Jews were forced to abandon their farms and homes and pursue other professions in the cities in order to survive. Large numbers of Jews left those countries and migrated to the United States in the late 19th and early 20th centuries and settled predominantly in urban areas, but the longing to return to the land did not die. Among those new arrivals were groups who came with a desire to rekindle the agricultural flame in their new country as yeomen. Some settled in agricultural colonies with support and assistance from the Jewish Agricultural Society and the Baron De Hirsch Fund. Others eschewed the communal setting for an entrepreneurial approach as independent farmers. Some joined the westward march and became pioneers of the untamed wilderness where they established ranches and cattle raising enterprises that are still in operation today. The Homestead Act of 1862 was a federal law that gave 160 acres of undeveloped land in the American West to any family head or person who was at least 21 years of age, provided he lived on the land for five years and built a house, or allowed the family head to buy the land for $1.25 per acre after six months. This legislation gave Jews the opportunity to try their hand at farming on land they owned in the developing West.

The population of Jewish farmers around the country grew rapidly during that period. Jews owned land in over 25 states and many needed advice and guidance in all aspects of operating a farm business. The National Farm School in Doylestown, Pennsylvania, was founded to educate and train Jewish youth in the techniques and special skills needed to function in a rapidly developing agricultural industry. In 1895, the colony at Woodbine established an agricultural school, named in honor of Baron Maurice de Hirsch, to educate the families of the New Jersey agricultural colonies.

Although some farming ventures failed due to economic, social, and political inequities, they set the stage for those who followed with enlightened ideas, new techniques, and a will to succeed. Following World War I there was a renewed interest in farming by immigrants fleeing the war-ravaged areas of Europe for a new life in the United States. As in earlier periods, a number of these individuals chose to live in rural communities and make their livelihoods on the farm. In addition, many soldiers returning home from the battlefields were attracted to agriculture or returned to their home farms in communities across the country. During the Great Depression of the 1930s, families seeking alternatives to the sweatshops and economic hardships of urban life decided to fulfill lifelong ambitions to farm. The 1940s and 1950s witnessed an additional upsurge in Jewish farmers when European Jews escaping the Nazis and the devastating effects of World War II arrived in America. The Jewish Agricultural Society and other Jewish agencies came forward to offer economic assistance and practical advice to the new immigrants who settled in farming villages and towns around the country.

During the 1960s and 1970s, America was in turmoil with political activism in civil rights, anti-war, and environmental movements challenging the status quo and seeking change. Among the options for young people during those years were "back to the land" movements that motivated progressive Jews seeking alternative lifestyles to leave their professions in urban areas for farm life, particularly in parts of New England and the Northeast. Although most of these new farmers had no previous agricultural experience, local farmers were available to help the newcomers adjust to farm life. They also received advice and assistance from agricultural extension services, short courses, and field days at their state colleges of agriculture. Many became successful and prosperous farmers who passed the farm along to second and third generation descendants that still manage the family's holdings.

The zeal for a Jewish farming experience has not abated. Even today, while the number of family farms is decreasing, replaced by large-scale agriculture, there are accounts of American Jews who embark on a second career in farming. In the education and service sectors one finds Jewish agricultural scientists, educators, extension agents, and employees of state and federal agencies contributing their expertise to making American agriculture the world's most abundant food production enterprise.

Jewish farmers have given America an intriguing legacy of determination, hard work, and a commitment to succeed and make an impact on a profession that is vital to human existence. This bibliography offers a view of that legacy as recorded in periodicals, newspapers, books, reports, conference papers, theses and dissertations, films, and miscellaneous documents such as speeches and essays. Sources listed in the following sections are arranged alphabetically by author:

Periodicals— This section presents articles in scholarly journals and popular magazines that show Jews in agricultural pursuits in settlements, individual farms, colleges of agriculture, and other positions where their love of the land is manifest.

Newspapers— This section cites accounts of Jewish farmers and farm communities as reported in urban dailies, regional newspapers, and Jewish newspapers, that kept readers informed about the role of Jewish farmers and their impact in their respective communities.

Books— This section includes books written by enlightened authors who brought the stories of Jewish farmers and their families to the reading public.

Book Sections— Some historians include a chapter or several pages in their books to fill in the historical gaps that other authors have left open; those chapters or sections are listed here.

Conferences, Symposia, Reunions— This section lists works that have been presented at conferences and symposia. Some of the gatherings were reunions or annual meetings of established Jewish farm groups who met to share their experiences and reminisce about their lives.

Dissertations, Theses, Essays— Students at colleges and universities carried out research to analyze the role of farming in Jewish life in America. These studies were the topics for master's theses and doctoral dissertations.

Reports from Government Agencies, Organizations, and Associations— This section lists reports issued by United States government agencies and private organizations that offered financial support and social services to farmers and their families. The reports give details about the purposes, activities, and accomplishments of the agencies and their impact on the lives of the farmers and their families.

Museum and Exhibit Catalogs— To give the general public an opportunity to learn more about this aspect of Jewish history, museums and research centers sponsored exhibits with accompanying catalogs featuring nostalgic recollections of Jewish farmers and the communities where they lived.

Film and Video—A few audiovisual documentaries were produced to bring the story of Jewish farmers to American homes. Those productions offer a chance to hear the farmers and their descendants talk about their lives.

Miscellaneous—This section comprises speeches, catalogs, guides to archival materials and special publications.

For those interested in locating specific works, title and subject indexes are included. Readers will learn a great many facts about this subject by browsing the cited works and reading the annotations.

The Bibliography

Periodicals

1. Abdill, George B. (1965). New Odessa: Douglas County's Russian Communal Colony. Part 1. *Umpqua Trapper*, 1(4): 10–14.

A commune called New Odessa was founded in 1882 by a group of Russian Jewish immigrants on 760 acres near Glendale, Oregon. The group had been members of *Am Olam* (The Eternal People) whose goal was to immigrate to America and establish an agricultural commune with a socialist life style. Citing excerpts from land and real estate records and excerpts from other articles about New Odessa, Abdill describes their journey to America, the choice of a site for the commune, and the individuals and organizations that assisted them. He explains the customs at a wedding held at the colony, the land, crops, and sources of income and leisure activities including lectures, discussions and debates that kept them intellectually involved. (See Part 2 in: *Umpqua Trapper* 2[1]:16–21, 1966.)

2. Abdill, George B. (1966). New Odessa: Douglas County's Russian Communal Colony. Part II. *Umpqua Trapper*, 2(1): 16–21.

Outline of the philosophy, moral codes, and spiritual brotherhood at the New Odessa Colony established in 1882 by a group of Russian immigrants. The colony was guided by Marxist doctrines and members were given assigned tasks on the farms, the garden, the timber, the kitchen, and household chores. Abdill provides a glimpse into the group's leisure activities such as music and cooking and preparation for a wedding of commune members. At its peak, New Odessa had 65 members including William Frey, a non-Jew and former Russian nobleman and strict vegetarian who was instrumental in the colony's development. Includes descriptions of other colonists and their backgrounds, skills, and contributions to the New Odessa experiment. The colony failed and a foreclosure returned the land to its original owner in 1888. Colonists left for other locations and new professions including physicians, lawyers, dentist, chemist, educator, and engineer. (See also Part I in: *Umpqua Trapper* 1[4]:10–14, 1965.)

3. Abrams, Jeanne (1986). Chasing an Elusive Dream: Charles Spivak and the Jewish Agricultural Settlement Movement in America. *Western States Jewish History*, 28(3): 204–217.

A biography of Charles David Spivak (1861–1972) who became involved with the Jewish colonization movement in his youth in Russia and headed the Am Olam group in his home town of Kremenschug. He came to America in 1882 with a dream to establish Jewish agricultural settlements. Spivak worked as a farmhand and teacher at the Alliance Colony in New Jersey; he wrote articles on farming for Jewish periodicals. Spivak studied medicine and became a doctor and moved to Denver, Colorado where he was active in the medical community. He helped establish the Jewish Consumptive Relief Society Hospital in Denver with a farm that provided a healthy work environment for tuberculosis patients. After settling in Denver, he visited the struggling Cotopaxi and Atwood colonies to find out why they were struggling to survive. He concluded that agricultural settlements fail due to internal dissension, poor management, and land unfit for cultivation.

4. Adler-Rudel, S. (1963). Moritz Baron Hirsch, Profile of a Great Philanthropist. *Leo Baeck Yearbook*, pp. 39–69.

A description of the life of Baron Maurice De Hirsch and his active involvement in philanthropic and humanitarian causes on behalf of Jews around the world. He is best known for supporting the establishment of agricultural colonies in Russia, Argentina, Palestine and the United States in the late 19th and early 20th centuries. In 1891, he established the Baron de Hirsch Fund to help immigrants from Russia with loans, transportation, training, language instruction, and education and settlement in agricultural colonies throughout the United States. Briefly covers the United States settlement efforts (pp. 43–45).

5. Allen, Lewis (1879). Founding of the City of Ararat on Grand Island — by Mordechai Noah. *Publications of the Buffalo Historical Society*, 1: 305–328.

Grand Island, New York, is located in the Niagara River about four miles below Lake Erie. The Island is nearly 12 miles long and varies from 3 to 7 miles in breadth with about 17,000 acres of fertile land. In this speech, presented to the Buffalo Historical Society on March 6, 1866, Allen gives details of the Island and its history including size, and topography, natural beauty, agricultural capabilities and its early inhabitants (squatters). He explains how Mordechai Noah became interested in Grand Island and his crusade to purchase the Island as a refuge for Jewish immigrants fleeing persecution. Allen discusses Noah's speeches, proclamations, and celebrations, purchase of the Island in 1824 and laying the Ararat cornerstone. Noah's plan was widely criticized by Jewish leaders and he was unable to attract large numbers of settlers to the Island. Allen points out that there are contradictory accounts about whether Noah ever saw or traveled to the Island. The idea was abandoned and the Island was sold in 1833. (For a definitive biography of Noah and his life including the Ararat project, see: Goldberg, Isaac. *Major Noah: American-Jewish Pi*oneer. Philadelphia, The Jewish Publication Society of America, 1936.)

6. Anonymous (1958 December 6). The Jewish Farmer. *America—National Catholic Weekly Review*, 100: 305.

A brief tribute to the work of Gabriel Davidson, General Manger of the Jewish Agricultural Society who died in November 1958. Mentions the activities of the Society and Davidson's dedication to Jewish farm settlement activities. Davidson supported the National Catholic Rural Life Conference in its efforts to maintain the family farm as an alternative to industrialized agriculture.

7. Anonymous (1953 August 1). Increase in Jewish Farmers. *America—National Catholic Weekly Review*, p. 430.

Brief summary of highlights of the 1952 annual report of the Jewish Agricultural Society. The number of Jewish farmers has increased since World War II with the largest number in New Jersey engaged in poultry farming. Jewish farmers can also be found in Connecticut, California, and Florida. Jewish farmers make a major contribution to agriculture with a gross annual production of $10,000 which represents only 14 percent of all farmers. Appeals should be made to recently arrived refugees to consider a life on the farm.

8. Anonymous (1938 March 14). Jewish Farmers. *Time*, 31(11): 68–69.

Summary of the 10th annual report of the Jewish Agricultural Society. Gives examples of successful Jewish farmers and prominent Jews who were raised on farms. Includes a brief history of the Society and its activities and observations regarding the religious and biblical inspirations that motivate Jews to choose farming as a profession.

9. Anonymous (1938). Clarion Utah Colony. *Utah Poultry Co-Operator*, 2(7): 12.

Benjamin Brown was a founder of the Jewish agricultural colony at Clarion, Utah and a founder of the Utah Poultry Producers Association. He left Utah in 1936 to establish a 1200 acre village called Jersey Homesteads near Hightstown, New Jersey. This is a summary of an article in *The Saturday Evening Post* about Jersey Homesteads and Brown's role as an organizer and leader of the village. (See also: Armstrong, Ralph. Four Million Dollar Village. *Saturday Evening Post* 210(32)5–7+, February 5, 1938.)

10. Anonymous (1936 September). Leaves from Our Album: Benjamin Brown. *Utah Poultry Co-Operator*, 1(1): 7.

A brief tribute to Benjamin Brown, a Jewish poultry farmer who began farming as a member of the Clarion (Utah) Colony and then became an independent poultry producer when the colony was disbanded. Brown was a founder of the Central Utah Poultry Exchange and the Utah Poultry Producers Association. He left Utah in 1936 to establish Jersey Homesteads, a 1200 acre village in Monmouth County, NJ.

11. Anonymous (1931). From Ghetto to Farm. *Literary Digest*, 109(4): 25.

Summary of the highlights of the 31st annual report of the Jewish Agricultural Society. In 1931, there was a Jewish farm population of 800,000 cultivating a total of about 1 million acres throughout the United States. In 1930, the Society loaned $6,550,000 to Jewish farmers and found farm employment for 17,060 men. The Society purchased a tract of land near New Brunswick, New Jersey to subdivide into 5 acre poultry and truck farms and settled a select group of workers who combine farm work with off-farm jobs. Ninety percent of the Jewish farmers intend to remain on the farm and 64 percent want their children to engage in some farm of agriculture.

12. Anonymous (1923 July). Our Jewish Farm Population 75,000. *World's Work*, 46: 241–243.

A summary of the accomplishments of American Jews in agriculture excerpted from a report of the Jewish Agricultural Society. There are 75,000 Jews living on farms totaling more than 1 million acres. This represents 2-1/2 percent of the 3 million Jews in the United States in 1923. The article points out that although Jews are mainly urban dwellers and a "trading people," they have been engaged in agricultural pursuits since biblical times and those who are farming in the United States are doing well. Includes illustrations of a chicken farmer and tobacco farmer.

13. Anonymous (1923 May 12). Jews as Tillers of the Soil. *Literary Digest*, 77(6): 31–32.

Summary of the 1922 report of the Jewish Agricultural Society (JAS). There are 75,000 Jewish farm families on 1 million acres of farmland with an estimated total value of $100,000,000. In the previous four years, farms were found for 468 applicants and 224 received loans from the Society. The majority of Jewish farmers are located in New York, Connecticut, Massachusetts and Pennsylvania with a income from tobacco, potatoes and poultry enterprises. The JAS Extension Department visited 1607 farmers and held 134 meetings, demonstrations, field days and extension schools. The report shows that Jews are returning to the honorable occupation of agriculture and are doing well as tillers of the soil.

14. Anonymous (1921 June). 50,000 Jews Farm a Million Acres in This Country. *Current Opinion*, 70(6): 837–838.

Describes the work of the Jewish Agricultural and Industrial Aid Society and its activities. The Society provides loans and assistance to settle Jews on farms. Its Farm Settlement Bureau has successfully settled 50,000 farmers on over a million acres in thirty seven states. The Farm Labor Bureau has secured jobs on farms for over 12,000 men.

15. Anonymous (1919 June). Hebrews Are Rapidly Flocking to Farms in This Country. *Current Opinion*, 66(6): 403.

A summary of the annual report of the Jewish Agricultural and Industrial Aid Society. The society has given 4849 loans totaling over $2 million dollars. The society also placed 9536 farm laborers in good paying, permanent jobs. Explains how loan applications are processed and the factors that are necessary for success in farming.

16. Anonymous (1914). How Jewish Farmers Cooperate in America. *Independent* 80(3447): 516, 518.

A news report about the four day convention of the Federation of Jewish Farmers of America meeting in New York City. The Federation, founded in 1909, has 63 chapters in 11 states and Canada. The aim of the Federation is to improve the social and material condition of Jewish farmers through cooperative efforts. With the assistance of the Jewish Agricultural and Industrial Aid Society, the Federation established a cooperative credit union, purchasing bureau, savings and loan association, and fire insurance company.

17. Anonymous (1913 November 29). Jewish Farmers in America. *Outlook*, 105: 686.

The Federation of Jewish Farmers is holding a convention in New York City to call attention to the large number of Jews engaged in farming and the issues that concern them. The organization has fifty branches in fourteen states with a membership over 500. Some of the activities of the Federation include establishment of an agricultural credit union, a fire insurance company, and adoption of labor saving equipment to help its members farm more effectively. Scholarships offered by The Jewish Aid Society to state agricultural schools has sparked an increase in the number of Jewish men and women studying agriculture. The College of Agriculture at Cornell has 150 Jewish students.

18. Anonymous (1913). Establishing a New Jerusalem in Sholam. *Olde Ulster*, 9(8): 225–231.

In August, 1837, five families arrived at the colony of Sholam in the town of Warwasing, New York in Ulster County. In May 1838, another group of colonists arrived bringing the total to about 15 families, including two Christian families. They purchased parcels of land and built houses to begin establishing an agricultural colony. Although they had common interests and shared a synagogue, all land was individually owned. The article describes some of the settlers, the cost of their land, and the reasons for joining the Sholam colony. Although they intended to make a living from farming, it was apparent from the outset that the colony would struggle to survive. The settlers were idealistic but had no prior knowledge of agriculture and they quickly found out that the soil was poor, the growing season was short, and they lacked farm experience. Some settlers managed to meet expenses by engaging in merchandising and peddling. Dissatisfaction, jealousy, and suspicion arose among the settlers which disrupted the unity of the colony. The experiment ended in 1840 when foreclosures were brought against some of the settlers and they began to leave the area. Some held out until 1847 and one settler, who paid in full for his land, remained in the area for 26 years. The article identifies some of the earlier settlers, land economics, and reasons for joining the colony. (For a followup letter regarding the settlers at Sholam Colony, see: De Sola Pool, The Jewish Colony of Sholam, Ulster County, New York, 1838. *Publications of the American Jewish Historical Society*, 23:178–179, 1915.)

19. Anonymous (1913). Jewish Farmer, Gentleman. *Independent.* 75(3376): 402.

Excerpt of a letter sent to the Farm Labor Bureau of the Jewish Agricultural and Industrial Aid Society from a Jewish farmer in Potsdam, New York. The letter praises Boris Flatow, a farmhand who was hired through the Society. Flatow had to return to Russia for family matters and the farmer regrets losing him. Following the excerpt is a brief description of how the society provides assistance to Jewish farmers with credit, counseling, education, and information services.

20. Anonymous (1912). The Jewish Colony at Sholam. *Olde Ulster*, 8(6): 161–166.

An account of the establishment of the agricultural settlement of Sholam at Wawarsing in Ulster County, New York in 1837. The Sholam residents were Russian immigrants from New York City that wanted to try to make a life in a mixed agricultural/industrial settlement. They were highly educated and had an interest in art and music. They bought several hundred acres of land in Ulster

County in December 1937 and began clearing the land to prepare it for farming. They built homes, a store, a synagogue, a museum, an art building, and two factories to manufacture quill pens and fur caps. Some members gained extra income as peddlers, others were traveling shoemakers and tailors. The economic crisis of 1837 and defaults on loans and mortgages forced the colony out of business by 1842. The settlers returned to New York and entered other professions.

21. Anonymous (1911). Director of Agricultural Station. *Jewish Charities*, 1(9): 8.

A biographical sketch of Jacob Lipman, who was appointed Director of the New Jersey Agricultural Experiment Station. Lipman was one of the settlers at the Woodbine colony, a graduate of the Baron De Hirsch Agricultural School, and a professor at Rutgers University, College of Agriculture.

22. Anonymous (1909 May 1). One Way Back to the Soil. *Survey*, 22: 162–163.

The National Farm School at Doylestown, PA offers two year instruction in agriculture and farming techniques to Jewish youth from urban areas. The course includes practical instruction in farm work, agricultural chemistry, veterinary science, farm hygiene, and practical mechanics. Instruction is carried out on the grounds of the school and three adjacent farms. Graduates of the school are employed by the USDA, state institutions, and independent farms.

23. Anonymous (1900). A Successful Exhibition at the Educational Alliance. *Menorah: A Monthly Magazine for the Jewish Home*, 41(4): 168–170.

Description of an exhibit held at the Educational Alliance in New York sponsored by the Baron De Hirsch Fund, the Jewish Agricultural and Industrial Aid Society and the Industrial Removal Office. Includes scenes from the Baron de Hirsch Agricultural School at Woodbine, NJ and the National Farm School at Doylestown, Pennsylvania, produce from farms at the Arpin, WI colony, and depictions of life in the Southern New Jersey colonies of Norma, Alliance, Carmel, Rosenhayn, Brotmanville, and Woodbine.

24. Anonymous (1885). A Wedding Among the Communistic Jews in Oregon. *Overland Monthly and Out West Magazine*, 6(36): 606–611.

Observations of a wedding which took place on the New Odessa Colony in Oregon in a roughly built community center. The unknown writer begins with observations about Jewish life in Russia and the despotism and brutality that forced them to emigrate to America. The writer then offers stereotypic views of Jews and their beliefs and life style. The detailed description of the wedding gives details about the bride and groom, sleeping arrangements, cooking, decorations, and guests.

25. Applebaum, Emanuel (1964). The Palestine Colony—An Agricultural Endeavor. *Michigan Jewish History*, 4(2): 14–17.

Story of the Jewish agricultural settlement of Bad Axe, Michigan, established in 1891 about 125 miles north of Detroit. By 1892 there were 16 farm units at Bad Axe, with a total population of 75 persons, the settlement was faced with a number of economic and social problems and all settlers left by 1900. Applebaum explains how the idea originated, who were the financial backers, economic problems, and reasons for failure. Settlers' lack of agricultural knowledge, crop failures, and pressure from creditors were major factors that led to the settlement's demise.

26. Armstrong, Ralph F. (1938 February 5). Four Million Dollar Village. *Saturday Evening Post*, 210(32): 5–7, 34, 36, 38, 39.

Story of the founding of Jersey Homesteads, a 1200 acre cooperative village in Hightstown, New Jersey in Monmouth County. It consists of one third farm land and orchard, two thirds houses and a garment factory and cannery. The colony began with 100 families (with plans to expand to 200 families), who were immigrants from Ukraine, Poland and other parts of Eastern Europe. Benjamin Brown, a former leader of the failed Clarion, Utah settlement, and founder of the Intermountain Farmers Association, left Utah and became the major activist for Jersey Homesteads. The project was supported by the Resettlement Administration (later renamed the Farm Security Administration) which spent $4 million to move the 100 families to the site. The factory and cannery provided one third of the revenue, with the farm and a general stores bringing in the other two thirds. Jersey Homesteads was plagued from the start with public opposition and economic troubles, agricultural uncertainty, government bureaucracy, and settler anxiety. Critics called it a communist idea and attempted to shut it down and expel the settlers. However, patience and determination by the settlers and Brown's belief in the project paid off when the first eight settlers moved into their homes in July 1936; the factory opened in August 1936 with 60 workers employed. Armstrong observes that "homesteaders and government have learned to expect almost anything at Jersey Homesteads."

27. Bent, Silas (1923). Three City-Bred Jews That the Farmer Trusts. *Outlook*, 12: 553–556.

Describes efforts by three Jewish millionaires Bernard Baruch, Eugene Meyer Jr., and Aaron Sapiro, bankers and Wall Street executives who advise and train farmers on techniques for running a successful farm business and overcoming the economic difficulties that they face. The three millionaires who made their fortunes in banking and Wall Street, offered their expertise to farmers, farm organizations, and States regarding consolidation, farm credit, and business management. They emphasize the formation of co-operatives that will enable farmers to grow, grade, and market their products. Consolidation and co-operation of the farm business will put the farmer on an equal basis with the buyers, processors, and related industries. Bent gives brief backgrounds of the millionaires and how they became interested in Jewish farmers.

28. Bermann, Simon (1859 September 8). Mortara — Catholicism and Judaism. Letter to the Editor. *Occident*, 17(24):142.

Continuation of a letter to the editor from a reader in Virginia discussing the Mortara Affair (the kidnapping and forced Baptism of a Jewish child) and other injustices against Jews in Europe. Bermann calls for the establishment of a Jewish Colonization Society in the United States dedicated to settling immigrants from those countries in agricultural colonies. He recommends establishment of a Jewish Agricultural Society in Virginia where forty or fifty thousand acres can probably be acquired for a moderate price. Virginia has a mild climate and soil conditions suitable for raising sheep and cattle, growing tobacco, and vineyard and orchard cultivation. Bermann believes that a successful colony would afford the Jews the importance and respect they are due in America and abroad.

29. Blumenthal, Helen E. (1982). The New Odessa Colony of Oregon, 1882–1886. *Western States Jewish Historical Quarterly*, 14(4): 321–332.

A group of young Jewish intellectuals from Odessa, Russia who were advocates of "Am Olam" (Eternal People) emigrated to the United States in 1882 and founded an agricultural colony on 760 acres of land near Glendale, Oregon which they called "New Odessa." It was planned as a Jewish Communist community with about 65 members at its height. Although only 180 acres were suited for agriculture they raised wheat, oats, barley, potatoes and vegetables; however their major source of income was selling cordwood to the railroad. Political discord, the poor quality of the land, and lack of privacy for married couples caused an exodus from the colony and it was abandoned after four years.

30. Bressler, David M. (1912). Work with Immigrants: Jewish Farmers on Government Homesteads. *Jewish Charities*, 2(7): 9–11.

Bressler visited over 350 Jewish farm families on settlements in North Dakota, South Dakota, Nebraska and Wyoming to determine what attracted them to farming and to observe how they are doing. He found that despite the hardships they face, the families demonstrated a pioneer spirit and a desire to stay on the land. Bressler notes the following obstacles that settlers faced: harsh climates, lack of nearby railroad facilities, children traveling great distances to school, lack of social contact with other Jews, crop failures or limited harvest, and large debts.

31. Bronstein, Zelda, and K. Kann (1977). Basha Singerman. *California Historical Quarterly*, 56(1): 20–33.

Oral interview with Basha Singerman (not her real name), an immigrant from Minsk who came to America with her husband, Shimon in 1915 and settled in the community of Jewish chicken farmers in Petaluma, California. Singerman discusses her life in Minsk, eleven months in Kenya after leaving Minsk, her arrival in the United States, and the decision to go to Petaluma and join the Jewish farmers. She recounts memories of life on the farm, ideological differences and activism in the community, death of Shimon (1942), and the economic hardships which led to bankruptcy and loss of the farm in 1966. Singerman tells anecdotes about the political activism, the reading and cultural club, the community center, and various members of the community. (This is a condensed version of a longer interview which appeared in Kann, Kenneth. *Comrades and Chicken Ranchers: The Story of a California Jewish Community.* Ithaca, Cornell University Press, 1933, pp. 15–27.)

32. Buttnick, Meta, and J.N. Eulenberg (2002). Republic, Washington: A Jewish Settlement in the Small Towns of Washington State, 1898–1936. *Western States Jewish History*, 34(4): 312–323.

At the turn of the Century, stories about gold in Republic, Washington brought many settlers to the area. In 1900, Republic was the sixth largest city in Washington. The gold frenzy ended by 1906 and the town's population dwindled to 800. News stories about the area in the Yiddish press, resulted in two waves of Jewish immigration to Republic by families interested in farming. At its peak, there were 25 Jewish families in Republic. During the first wave, 1906–1910, six families who were unprepared for agriculture and poorly trained, defaulted on loans and sold the land and moved away. The second wave occurred in 1914 when a group of colonists from Glace Bay, Nova Scotia attempted to set up a farming and stock raising company. The colony lasted only a few years because the land was not suitable for farming and the long hours and hard work were too difficult. The colony was eventually abandoned and the settlers left.

33. Clar, Reva (1984). Agricultural Gleanings from San Francisco's "Weekly Gleaner." *Western States Jewish History*, 17(1): 52–62.

Rabbi Julius Eckman (1805–1874), the first ordained Rabbi in the Far West, was the leader of San Francisco's Congregation Emanu-El and founder of the city's first religious school. He was also founder and editor of the *Weekly Gleaner* a widely distributed newspaper that included writings on a range of subjects, both religious and secular. Eckman wrote about the agricultural potential of California and the rewards of agriculture as an occupation for Jews seeking alternatives to careers in business. Clar offers a summary of some of those writings including excerpts about successful Jewish farmers who contributed to the State's agricultural development.

34. Clar, Reva (1984). Tillie Lewis: California's Agricultural Industrialist. *Western States Jewish History*, 16(2): 139–155.

Story of the life and business career of Tillie Lewis (1901–1977) who established a cannery business in Stockton, California and became a prominent food packer and shipper. Working with California farmers, Lewis was responsible for development, growth and processing of the pomodoro tomato and tomato products. She helped developed practical uses for asparagus by-products. She developed a procedure for packaging combat rations during the Korean war. She packaged foods for wholesale groceries like A & P and other companies and created a line of diet and health foods.

35. Cone, G. Herbert (1903). New Matter Related to Mordecai M. Noah. *Publications of the American Jewish Historical Society*, 11: 131–137.

Mordecai M. Noah was a lawyer and journalist whose family traced roots back to the revolutionary period. He is best remembered for his plan to establish a Jewish colony on Grand Island on the Niagara River near Buffalo, New York. He petitioned the New York State Assembly to consider his scheme and grant Grand Island to him. This paper presents excerpts from the petition, correspondence, newspaper articles, and letters from supporters and critics of Noah.

36. Cooley, Everett L. (1968 Spring). Clarion, Utah: Jewish Colony in Zion. *Utah Historical Quarterly*, 36(2): 114–131.

In 1911, a group of 150 Jews from New York and Philadelphia, selected a site in Clarion, Utah to establish an agricultural colony. They formed the Jewish Agricultural and Colonial Association with assistance from the Jewish Agricultural Aid Society and the National Farm School. They acquired 6,085 acres on arid lands near Gunnison, Utah and plans were made to settle 52 families by Spring 1913. Although the settlers were optimistic about the colony's success, they encountered serious problems from the outset. Economic difficulties, lack of experience, unproductive land and other problems arose and the colony was disbanded by 1916. Benjamin Brown, the leader of the colony, and a few other settlers continued to farm in Utah and established a thriving poultry industry which was incorporated in 1923 as the Intermountain Farmers Association. The farmers raised chickens and eggs and opened a grading plant in Gunnison and eventually expanded the market to California. By 1967, the Association had gross revenues of over $9 million. Although the attempts to form a colony had failed, a successful legacy remains.

37. Cowan, Max P. (1977). Memories of the Jewish Farmers and Ranchers of Colorado. *Western States Jewish Historical Quarterly*, 9(3): 218–225.

Cowan was a livestock dealer, cattle rancher, and feedlot operator. He raised cattle and sheep in Western states and operated feedlots in California and Texas. In this excerpt from his memoirs, he gives personal recollections of successful Jewish farmers and ranchers that he was personally acquainted with in Colorado. They came to the area as young immigrants during the late 19th and early 20th century after trying other ways to make a living. They settled on the land, raised their families and made their livelihoods in agriculture. Their offspring were not interested in a life on the farm and left to find other occupations elsewhere.

38. Cowen, Alfrida D. (1917). Moses Elias Levy's Agricultural Colony in Florida. *Publications of the American Jewish Historical Society*, 25: 132–134.

Details of Moses Elias Levy's plan to establish an Jewish agricultural colony on 36, 000 acres of plantation land he owned in Alachua County, Florida. The colony was established in 1820 at great expense. Although many potential settlers declined to participate, several wealthy families from New York, New Jersey and Europe came to Florida to settle in the colony. By 1823 there were 50 persons settled in 25 houses. Levy cleared roads, built a sawmill and ultimately established 3 plantations.

39. Dan, Joseph (1994 March). A Jew Outstanding in His Field. *B'nai B'rith International Jewish Monthly*, 108(5): 33.

Story of Gary Goldberg, a Bronx raised New Yorker who farms 400 acres of corn and soybeans in Nebraska. Goldberg went to college in Nebraska where he studied psychology, but he became interested in farming after marrying a farmer's daughter from Nebraska and worked on his father in law's farm. He bought his own farm and began raising hogs and later switched to corn and soybeans. Goldberg is president of the American Corn Growers Association and often lobbies Congress on farm bills and trade agreements. Goldberg states that he loves farming because it gives him a feeling of accomplishment. He wants to use his experience to convince urban residents to appreciate where their food comes from.

40. Davidson, Gabriel (1944 November 17). His Ancestral Calling. *Commonweal*, 41(5): 120–122.

Davidson, President and General Manager of the Jewish Agricultural Society, claims that Jews have always been involved in agriculture and the "modern farmward movement" among Jews is a return to an ancestral calling. He provides brief examples of successful Jewish farmers, agricultural scientists, and educators, who made important contributions to American agriculture.

41. Davidson, Gabriel (1935). The Jew in Agriculture in the United States. *American Jewish Year Book*, 37: 99–134. A follow up to Leonard Robinson's 1935 "Agricultural Activities of the Jews in America" (*American Jewish Year Book* 14:1–115, 1912) which examined the Jewish farm scene from 1820 to 1912.

In this update, Davidson presents facts, figures, and statistics about Jewish farmers in the 1930s. There are over 109,000 Jews living in rural areas with many of them engaged in farming. The chief enterprises are dairy, poultry, vegetable, and fruit production on farms that vary widely in acreage and soil types. To illustrate the variation, Davidson briefly describes Jewish communal life in Connecticut, New York, New Jersey, Ohio, and Michigan. He offers observations on how the agro-industrial development model (combining small industry and off farm jobs with agriculture) and the activities of the Jewish agricultural agencies and the Federation of Jewish Farmers that gave farmers the financial, technical, and educational support they needed to remain in business. Davidson lists the credentials of Jews on the staffs of agricultural schools and colleges, experiment stations, extension service, state and Federal bureaus and scientific research agencies. He concludes with a summary of a sociological study of Jewish farm life conducted in 1935 by the Jewish Agricultural Society showing ages, diversity of occupations and skills, capital, types of farms, number of years on the farm, relations with non-Jewish neighbors, and observations on what is necessary for success in farming.

42. Davidson, Gabriel (1925). The Palestine Colony in Michigan: An Adventure in Colonization. *Publications of the American Jewish Historical Society*, 29: 61–74.

The story of the Palestine Colony in the thumb district of Michigan in a small village called Bad Axe, about 125 miles North of Detroit and fifty miles east of Bay City. Russian immigrant Jews living in Bay City during the depression of the 1890s, working mainly as peddlers purchased land near Bad Axe and began to establish a settlement in 1891. Twelve families (about 57 people) bought parcels of land and began clearing it, building houses and purchasing equipment. They received financial support from Temple Beth El Hebrew Relief Society in Detroit and the Baron De Hirsch Fund. Davidson describes what it was like for the settlers who struggled to make it work but were faced with difficulties from the start. Although there were some good years, lack of farming experience, perpetual debt, infertile land, crop failures, inadequate housing, and petty disputes made it difficult for Palestine Colony to succeed. Settlers began leaving in 1899 and by 1900, it was abandoned. (Also published slightly condensed in: Davidson, Gabriel, *Our Jewish Farmers and the Story of the Jewish Agricultural Society*. New York, L.B. Fischer, 1943, pp. 234–249.)

43. Davidson, Gabriel, and Edward A. Goodwin (1928 May). A Unique Agricultural Colony. *Reflex*, 2(5): 80–86.

A history of the New Odessa Jewish Agricultural Communal colony in Oregon. The roots of the commune idea can be traced back to the Am Olam movement in Odessa, Russia. Am Olam was a movement started in Russia with the belief that farming was a noble profession and should be exemplified by establishing agricultural colonies on a socialist model in Palestine and America. Between 1881 and 1882, four groups of Am Olam migrated to the United States and members from those groups established the Oregon Colony. Davidson and Goodwin explain the steps leading to the development of New Odessa. They present details about the land and soil conditions, work routines, living conditions and the differing ideologies of the members of the colony. At its height, the colony had about 65 people. There were some marriages between members, the birth of a child, and the death of a member. The colony lasted about two years and then began to disintegrate due to bickering and irreconcilable differences. The authors point out that geographic isolation and intellectual stagnation and a desire for books and careers also contributed to the closing of Odessa colony. Members left the colony for cities like San Francisco and New York where they entered other non-farm jobs and businesses. The New York group maintained their socialist lifestyle and lived in a household with shared tasks. That arrangement fell apart after five years. Others entered careers in medicine, law, pharmacy, dentistry, chemistry, engineering and education. (Also published in: Davidson, Gabriel, Our Jewish farmers and the story of the Jewish Agricultural Society. New York, L.B. Fischer, 1943, pp. 226–233.)

44. Davis, Charles K., I.M. Wise and M.H. Marks (1965). A Colony in Kansas—1882. *American Jewish Archives*, 17(2): 114–139.

The Hebrew Union Agricultural Society (HUAS) in Cincinnati attempted to settle Russian Jewish immigrants on an agricultural colony in Beersheba, Kansas in 1882. Charles K. Davis joined the Beersheba colonists to assist in the settlement and training.. The "Editor's Introduction" (pp. 114–115) offers some thoughts on agricultural settlement for newly arrived Jewish immigrants. In reprints from issues of *American Israelite (1882–1883)*, Davis gives his observations of the Beersheba experiment with excerpts from his personal diary. "A Humble Plea for a Russian Colony," by Isaac M. Wise (p.116–118) urges the Russian Emigrants' Aid Society and other supporters of the settlement initiative to raise funds and assist the colony to survive and prosper. In "Beersheba Colony—An Assured Success" (pp. 134–139) M.H. Marks, President of the Hebrew Union Agricultural Society (HUAS), describes his visit to the Beersheba Colony and the notable spirit and independence of the settlers. He believes that Jews can be successful farmers and offers continued support of HUAS.

45. De Sola Pool, David (1915). Jewish Colony of Sholam, Ulster County, New York, 1838. *Proceedings of the American Jewish Historical Society*, 23: 178–179.

A letter to the editor responding to an article about the Jewish Colony at Sholam that appeared in *Olde Ulster* (June 1912). The article stated that eleven Jews purchased land at Sholam and their names are on record. De Sola Pool describes one of the colonists named Marcus Van Gelderen. Van Gelderen, born in 1798 in Holland, came to the United States in 1816 at age 24 and worked as a *schochet* and *bodek* (supervisor of ritual slaughter according to Jewish law) at Shearith Israel in New York before joining Sholam. The colony failed and went out of existence about 1842. There are no details of his life from 1842 through 1859. From 1860 until his death in August 1871 Van Gelderen served again as *shochet* and *bodek* at Shearith Israel. Another detail concerning the colony at Sholam is a letter signed by Moses Cohen and Van Gelderen sent to the Board of Trustees of Shearith Israel in April 1838 asking for aid to build a house of worship at Sholam. The Board postponed action on the request. (See also: Anonymous. The Jewish Colony at Sholam, Ulster County. *Olde Ulster* 8(6):161–167, 1912.)

46. Deutsch, Gotthard (1982). David Lubin: A Remarkable Jew. *Western States Jewish Historical Quarterly*, 14(4): 316–320.

A biographical sketch of David Lubin, merchant and department store founder who believed in promoting agriculture and rural life among the Jews. He established the agricultural colony at Orangevale, California and proposed the establishment of rural credit unions and cooperation among agricultural organizations.

47. Douglas, Donald M. (1993). Forgotten Zions: The Jewish Agricultural Colonies in Kansas in the 1880s. *Kansas History*, 16(2): 108–119.

An overview of seven Jewish agricultural colonies established in the 1880s on the plains of southwestern Kansas mainly by Russian immigrants and some

Romanian and Hungarian immigrants. The colonies were sponsored by organizations or individuals, some reflected the ideas of the Am Olam organization, and others were pioneering efforts under the homestead programs. The seven colonies, Beersheba, Montefiore, Lasker, Hebron, Gilead, Touro and Leeser failed within 2–6 years due to bad soil conditions and poor choice of land, crop failures, deflated agricultural prices, isolation from nearby markets, and harsh climate. Many left the areas to settle elsewhere, others abandoned agriculture and returned to urban occupations.

48. Dubrovsky, Gertrude W. (1989). Der Yiddisher Farmer. *Reform Judaism*, 18(1): 5–7.

A brief history of the Jewish farming community of Farmingdale, New Jersey. The community comprised settlers from four waves of immigration: Eastern European immigrants in the 1920s, German Jews escaping Hitler in the 1940s, American born professionals who had served in WWII in the 1940s and Holocaust survivors from Eastern Europe in the 1950s. The settlers had progressive views and the community center was a place to meet and discuss the condition of the world and the plight of Jews. The community reached its height in 1953 and then declined due to changes in the farm economy, overproduction and withdrawal of price supports, and political polarization among the settlers.

49. Dubrovsky, Gertrude W. (1988). In Search of America: Lexington, KY and Farmingdale, NJ. Midstream, 34(7): 29–33.

Dubrovsky compares life in two Jewish farming communities, Lexington, Kentucky and Farmingdale, New Jersey as reflected in literary works and oral interviews and archival documents. Israel J. Schwartz, who moved to Lexington in 1916, wrote about his experiences in *Kentucky* (University of Alabama Press, 1990), a book of nine poems. An archive known as the Farmingdale Collection (available at the YIVO Institute, New York) contains transcripts of oral interviews, documents and photographs of 120 Jewish farmers from Farmingdale. (This archive was the source for Dubrovsky's book *The Land Was Theirs: Jewish Farmers in the Garden State* [University of Alabama Press, 1992].) Dubrovsky examines the visions, beliefs and attitudes of the farmers and their families in Lexington and Farmingdale regarding stereotypes, relations with Christian neighbors, and farm life in both communities.

50. Dubrovsky, Gertrude W. (1981). Growing Up in Farmingdale. *American Jewish History*, 71(2): 239–255.

Dubrovsky reminisces about her life as the daughter of Polish immigrants growing up on a farm in Farmingdale, New Jersey in the 1920s and 1930s. She warmly recounts her experiences on the farm, school, and community, and impressions of her parents, teachers, and friends.

51. Dubrovsky, Gertrude W. (1977). Farmingdale, New Jersey: A Jewish Farm Community. *American Jewish Historical Quarterly*, 66(4): 485–497.

In 1919 the first Jewish farmer moved to Farmingdale, New Jersey. He was a Russian immigrant who was persuaded by the Jewish Agricultural Society to buy a farm. Others followed and within 10 years there were 50 Jewish farm families in Farmingdale. Many farmers also took in summer boarders to supplement their income. Dubrovsky's family moved to Farmingdale in 1932 and engaged in poultry farming. She discusses her life in the community and the benefits and problems of being a Jewish farmer in a predominantly Christian town. She describes the businesses, schools, economic, social, and cultural situations, and anti–Semitism in the area. The Jewish Agricultural Society provided a loan for building a Jewish Community Center in 1930 and a Yiddish school affiliated with the Sholem Aleichem Folk Institute provided instruction to the children. Farmingdale's Jewish character underwent changes in the 1940s when younger, more progressive Jews moved into the area and in the 1950's when another group arriving from the displaced persons camps of Europe arrived with expectations of a better life. Changing economic and social patterns gradually changed the community as the residents left the area seeking other professions. By the early 1970s, all the Jewish farmers were gone from Farmingdale.

52. Durland, Kellogg (1908 April). Immigrants on the Land: II. Jewish Farmers. *Chautauquan* 50(2): 248–265.

Durland observes that Jews can be successful in agriculture and offers a rationale for encouraging Jewish immigrants to settle in agricultural colonies. He describes the colonies of Alliance, Norma, Brotmanville, Carmel, Rosenhayn, and Woodbine and the economic, social, and cultural benefits that contributed to the colonies' successes. Includes photos of activities and buildings at Woodbine.

53. Eisenberg, Ellen M. (1993). The Limits of Gender Equality in 19th Century American Jewish Colonies. *Communal Societies*, 13: 71–83.

The agricultural colonies that were established by the Am Olam movement were committed to gender equality and tried to follow those tenets in their daily activities and work assignments. Eisenberg explores the contradictions between the ideological commitment to gender equality and the realities of the time that worked against it. Forces such as legal requirements that only men can be signatories to an agreement, economic forces, and pressure by sponsors and non-ideological members, limited the colonists ability to function in a fully egalitarian environment. Other factors including wage differentials, beliefs about women's fitness for hard work, 19th century Jewish values, and a male dominated larger society posed additional barriers that worked against the achievement of totally egalitarian colonies.

54. Eisenberg, Ellen M. (1992). Immigrant Origins and Sponsor Policies: Sources of Change in South Jersey Jewish Colonies. *Journal of American Ethnic History*, 11(3): 27–40.

The *Am Olam* movement established in the Pale of Settlement in Russia focused on establishing agricultural colonies in the United States to provide immigrants a productive life on the land as farmers. The colonies established in Salem and Cumberland counties, New Jersey in the late 18th and early 19th centuries were inhabited originally by Russian immigrants who had come from the Am Olam movement with an orientation to life in a cooperative, socialist style agricultural settlement. Over time, new immigrants who wanted an American way of life joined the settlements and forced a shift to mixed agricultural-industrial orientation geared toward individual ownership and entrepreneurial skills. The article examines the motivations that led to the founding of the settlements and the requirements of sponsoring organizations, and differing immigrant needs that changed the settlement patterns and enabled them to expand and survive.

55. Eno, R. D. (1989). Into the Woods. *Reform Judaism*, 18(1): 8–9.

Eno reminisces about the back to the land movement in New England in the 1960s and 1970s by Jews who left the cities for rural life styles. He describes their lives as yeoman farmers and the renewal of Jewish life in rural Maine, Vermont, New Hampshire, and southern Quebec.

56. Fairbanks, George R. (1940 January). Moses Elias Levy Yulee. *Florida Historical Quarterly*, 18(3): 165–167.

A brief biography of Moses Elias Levy Yulee (he dropped Yulee when running for Congress), a Florida businessman who attempted to establish a plantation in Florida as a refuge for the Jews. This sketch, written by a judge and friend of Levy, gives details about Levy's character, relations with father and brother, and his business acumen.

57. Feld, Lipman Goldman (1970). New Light on the Lost Jewish Colony of Beersheba, Kansas, 1882–1886. *American Jewish Historical Quarterly*, 60(2): 159–168.

There were eight attempts to establish Jewish agricultural colonies in Kansas during the 1880's. The first and most successful for a short period was the Beersheba colony established in 1882 under sponsorship of the Hebrew Union Agricultural Society (HUAS) in Cincinnati. About sixty Russian immigrant Jews settled on the land located 22 miles northeast of Cimarron on the Santa Fe Trail. The colony survived until 1886 when it disbanded due to lack of farm experience and agricultural knowledge, drought and semi-arid conditions, poor planning, and strained relations between HUAS directors and settlers.

58. Fields, Harvey J. (2002). Joseph Nudelman: The Dakota Years, 1844–1934. *Western States Jewish History*, 35(1): 9–19.

Joseph Nudelman was born in a small town near Odessa, Russia and migrated to the United States in 1881. He was a member of Am Olam and a socialist who wanted to follow the Am Olam dream of creating a farming settlement in the United States. Fields, Nudelman's great-great grandson, tells the story of Nudelman's life in the United States and the agricultural ventures he pursued. Nudelman was one of the original 22 Russian Jewish families that settled in the Painted Woods Colony in the Dakota Territory. When Painted Woods failed in 1885 many of the settlers left and returned to urban centers to pursue other professions but Nudelman remained on his land determined to succeed. He sold out in 1892 and moved to Portland, Oregon. After a year in Portland, Nudelman began to move his family to other locations over the next few years to try farming in other locations. He acquired land near Porterville, Folsom, San Reymon, and Heywood, California. A short time later he moved his family again to a ranch near Carson City, Nevada and finally to a ranch near Genoa, Nevada. In 1902, the Ranch was hit by a storm that destroyed the land and buildings. Nudelman, then 58 years old, decided to give up farming and returned with his family to Portland where he opened a meat market. Nudelman died shortly before his 91st birthday in August, 1935. (For a more detailed account of Nudelman's life and family see: Nudelman, Joseph R. *The Family of Joseph Nudelman.* Published by Eugene Nudelman, 1969. For more details on the Portland years, see: Lowenstein, Steven. *The Jews of Oregon, 1850–1970.* Portland, Jewish Historical Society of Oregon, 1987).

59. Fierman, Floyd S. (1982). Sol Floersheim: Jewish Rancher and Merchant in Northern New Mexico. *Western States Jewish Historical Quarterly*, 14(4): 291–301.

Biography of Sol Floersheim who was born in Germany in 1856 and migrated to the United States in 1879. After working for a short time in a Matzo factory in New York he moved to Trinidad, Colorado and went to work for a merchant family in that city. He later made his way to Las Vegas, New Mexico where he worked for eight years with the Charles Ilfeld Company, a wool and sheep trader. In 1897, he purchased a ranch called "Jaritas Ranch" and a mercantile business in Springer, New Mexico and a store in Roy, NM. Floersheim raised sheep and acquired more land eventually reaching 63,000 acres. Jaritas was raising 88,000 sheep in its early years. Floersheim acquired other partnerships in general merchandise, lumber, and mills. He died in 1946 and the ranch operation was taken over by his son and grandson who received degrees from Colorado State University in animal husbandry. It was still in operation in 1982, mainly as a cattle ranch. Provides details about Floersheim's life and family, his business dealings, and religious activities.

60. Frankenstein, Ruth J. (1990). Herman J. Levine, Manager of the Jewish Agricultural Society. *American Jewish Archives*, 42(2): 167–181.

A biography of Herman J. Levine, manager of the Jewish Agricultural Society. Frankenstein, his daughter, warmly describes her father's devotion to helping Jewish farmers get established and pursue a life in agriculture. She discusses her dad's early life, his desire to get American Jews established in farming, his fund drives, community activities, interfaith initiatives, and other roles as an advocate for the Jewish agricultural movement.

61. Freedman, Morris (1950 September). The New Farmers of Lakewood: A Jewish Community on the American Soil. *Commentary*, 10(3): 234–244.

Description of a community of Jewish chicken farmers in Lakewood, New Jersey about 60 miles from New York City. The community began to grow in the late 1930's when Jews in New York with a wide range of occupations and skills, decided to leave the city for "security and a pleasanter life" on the farm. A number of Russian Jewish refugees joined the settlement in the early 1940s. This unique community, had a cooperative nursery school, a little theater troupe, two large farmers' cooperatives, and a weekly discussion group. Freedman visited some of the families and observed their work style and listened to their stories about farm life in Lakewood. Though largely secular and non-observant, Lakewood farmers were aware of their Jewish heritage and did not have much contact with the Gentile farmers. Freedman describes how the community developed and revolutionized poultry production in the area. The community is committed to the welfare of its children and supports the nursery school programs, and has an active program of educational lectures, discussion groups, and parent teacher associations. There is a sense of independence among the farmers and a cooperative spirit when pitching in is necessary. Although the community appears to be happy, Freedman indicates there are tensions among farmers which creates gossip and rumors. Some of the marriages are not happy and the community seems preoccupied with psychiatry and mental health issues. Freedman concludes by stating that "the members of the community find it a useful and desirable social organism." (Also published in: Cohen, Eliot E. Commentary on the American scene: portraits of Jewish life in America. New York, Alfred A. Knopf, 1953, pp. 127–146.)

62. Freund, Charles J. (1912 January). Significance of the Jewish Farm Colony at Clarion, Utah. *Improvement Era*, 16(2): 106–111.

This is in response to a letter from the President of the Mormon Church praising the establishment of the Jewish agricultural colony at Clarion Utah and offering fellowship and help. Rabbi Freund describes the colony at Clarion and how it originated. He points to the role of agriculture in ancient Judaism and the fact that this colony is a return to that tradition. Most members of the Clarion Colony are Russian immigrants. Freund states that those who wanted a life in farming but were denied that opportunity in Russia can be part of the American melting pot and choose the occupation of farming not only to make a living but to do their part to "reclaim the earth for the service of man."

63. Friedman, Richard (1990 Winter). Some Jewish Farmers (Who Happen to be My Ancestors). *Memories of the Jewish Midwest*, 6: 35–42.

Friedman reminisces about his paternal grandparents, Sam and Fanny, who migrated to the United States from Lithuania in 1887 and settled in Iowa. In 1900, they acquired a 160 acre homestead in Gregory County, South Dakota and farmed the land until 1947 when it was sold. They lived on the farm during the summers and returned to Iowa after the harvest. Friedman recalls memories of summers spent on the farm and his grandfather's role in overseeing the farm operation and his relationships with other farmers in the area.

64. Geffen, Joel S. (1971). Jewish Agricultural Colonies as Reported in the Pages of the Russian Hebrew Press, *Ha-Melitz* and *Ha-Yom*: Annotated Documentary. *American Jewish Historical Quarterly*, 60(4): 355–382.

During the period of political persecution of Jews in Russia in the 1880s, the Am Olam and Bilu movements encouraged emigration of Jews to the United States to make their livelihoods as farmers in socialist or collective settlements. About 25 collective settlements were established and reports of the colonization efforts appeared in the Russian Hebrew Press. Contains excerpts in English from letters and articles in *Ha-Melitz* and *Ha Yom* reporting on colonies in Sicily Island, Louisiana; Alliance and Vineland, New Jersey; New Odessa, Oregon, and Winnipeg, Canada.

65. Gerard, Helene (1986). Yankees in Yarmulkes: Small-Town Jewish Life in Eastern Long Island. *American Jewish Archives*, 38(1): 22–56.

Gerard explores the social, cultural, and religious life of German Jewish immigrants who settled in Eastern Long Island between 1840 and 1870. Data was gleaned from over one hundred oral interviews done in sixteen villages comprising fishing and farming villages, factory towns and commercial centers in Breslau, Sag Harbor, Greenport, Bellport, Patchogue, Riverhead, and Setauket. Gerard discusses the hardships and benefits of rural life, community relationships, upward mobility and changes over time. The article concludes with a brief comparison to Jewish life in the region in the 1980's.

66. Ginsberg, Louis (1958). The Jewish Colony at Waterview. *Virginia Magazine of History and Biography*, 66(4): 459–462.

The only recorded Jewish agricultural settlement in Virginia known as "Inglewood" was established at Waterview in Middlesex County in 1882–83. The land was purchased by Joseph Friedenwald, a Virginia landowner. About 10 or 15 Polish immigrants from Boston, settled in Inglewood in two frame cottages. The Russian Aid Committee of Baltimore sent funds to purchase implements, seeds and stock. The colony was disbanded after six years because of illnesses, deaths, lack of farming knowledge, weather, and crowded living conditions. The remaining residents returned to Baltimore. Other attempts to establish settlement of Jewish farmers in Virginia were not successful. In 1876 Friedenwald purchased 286 acres on the Rappahannock River and 414 adjacent acres in 1882. It is not known if these were purchased for personal use or agricultural colonies.

67. Goering, Violet, and O.J. Goering (1984). The Agricultural Communes of the Am Olam. *Communal Societies*, 4: 74–86.

The Am Olam was group of Russian émigrés who came to the United States to establish Agricultural colonies patterned on the ideas of Leo Tolstoy, Robert Owen and Charles Fourier. This article focuses on four of those colonies: Sicily Island, Louisiana; Cremieux and Bethlehem Yehuda, Dakota Territory, and New Odessa, Oregon. Georing explains how the colonies were organized, some of the important members who had an impact on their colonies and the differences in ideology. Political differences, ideological conflicts, isolation, inclement weather, bad soil and land conditions, and debts and foreclosures eventually led to the disbanding of the colonies and the end of the Am Olam movement. After leaving the colonies, some settlers got involved in agricultural education, marketing cooperatives and farm publications. Others became active in politics and the labor movement. Goering points out that their ultimate goals were achieved even if for just a short period.

68. Goering, Violet, and O.J. Goering (1982). Jewish Farmers in South Dakota — the Am Olam. *South Dakota History*, 12(4): 232–247.

Discusses efforts to settle Am Olam members in two areas in South Dakota in the early 1880's. Am Olam were well-educated, secularized Russian Jewish youth, who emigrated to the United States with the express intent of setting up utopian agricultural communities. The authors trace the social and political background of the Am Olam, founded in Odessa in 1881 by two Jewish teachers, Monie Bokal and Moses Herder, and examines their ideologies with particular reference to the notion of "a return to nature and to productive labor on the land," which reflected the philosophical views of the late 18th and 19th centuries. Goering describes two colonies, one at Cremieux and one at Bethlehem Yehuda, during 1882–85, and reveals the sharp contrasts in the ideology and social structure of the two communities. Both colonies failed and were disbanded in 1885 for a number of reasons which included rigors of frontier life, factionalism, lack of farming experience, repeated crop failures, and disagreements about task assignments and division of labor.

69. Gold, David M. (1981). Jewish Agriculture in the Catskills, 1900–1920. *Agricultural History*, 55(1): 31–49.

The New York State counties of Sullivan and Ulster had about 161 Jewish farmers in 1904 which grew to over one thousand by 1912. Although they received financial support from the Baron De Hirsch Fund and the Jewish Agricultural Society, the farmers began to host paying boarders to increase their farm income. This was the impetus for the establishment of resorts in later years. The Catskills developed into a major vacation area for New York City residents seeking an escape from hot summers. They also established cooperative ventures to help them survive such as an agricultural credit union, savings and loan association, fire insurance, feed cooperative, and a rural sanitation department. The agricultural depression after WWI, the growth of the resorts, and government programs restricting credit and farm subsidies, led to a shift away from farming and loss of the Jewish character of agriculture in the two counties by 1920.

70. Goldberg, Robert Alan (1989). Building Zions: A Conceptual Framework. *Utah Historical Quarterly*, 57(2): 165–179.

Goldberg explores the concept of colonization and the factors that influence and impact upon settlement programs. He compares three settlement projects — the Jewish settlement in Clarion, Utah in 1911 which lasted until 1916; Mormon settlement in the Great Basin region comprising 500 settlements in seven states with only 69 failures, and Jewish settlement in Palestine in the 1830s and the 1900s and the establishment of successful *moshavot* (cooperative farms). To explain why settlement efforts either succeeded or failed, Goldberg focuses on the effect of five interrelated variables — farming experience, environmental conditions, capital availability, colonists' morale, and the existence of alternatives that offered economic security.

71. Goldblatt, Roy (1991). From Ghetto to Ghetto... and Maybe a Farm, Too. *American Studies in Scandinavia*, 23(1): 1–14.

From the 1880s through the early 1900s, Jewish immigrants from the ghettos of Eastern Europe came to the cities of the Northeast. They settled in all the major cities but New York was the major port of call. Many lived in crowded tenements and worked long hours in sweatshop industries. Some moved to less crowded areas with better living conditions and some left the cities seeking a better life in suburban communities. Small groups of immigrants were interested in agriculture and joined the move to establish farming communities. With assistance from the Baron De Hirsch Fund, Jewish colonization Association, Jewish Agricultural and Industrial Aid Society and the Hebrew Union Agricultural Society, the immigrants moved to settlements in Louisiana, Arkansas, South Dakota, Oregon, New Jersey and New York. Some colonies failed in a short time due to lack of farming skills, economic hardships, harsh weather, and political factionalism. Others such as Alliance and Vineland (New Jersey) colonies and the independent family farmers of Ulster and Sullivan counties (New York) remained in existence through the 1960s. Their success is attributed to nearness to large markets, proximity to large Jewish communities and acquisition of additional income from taking in borders or renting vacation bungalows and adding small industries for a mixed economic base. Retirements and lack of interest by their offspring who left the farms to pursue other careers led to the end of those Jewish farm enterprises in the two States.

72. Goldsmith, Steven R. (1978). Sam Hamburg: World's Foremost Jewish Farmer. *Western States Jewish Historical Quarterly*, 10(4): 330–342.

Biography of Sam Hamburg (1898–1976) who was considered one of California's most successful and innovative farmers. Hamburg was born in Russia and immigrated Palestine at age fifteen to get a basic education. He came to the United States in 1920 where he enrolled at University of California/Berkeley and received his B.S in agriculture in 1924. Hamburg worked for a short time as a tenant farmer. In 1932 he acquired land in the Los Baños area with the financial support of friends. His farm became a prosperous 8000 acre enterprise raising, melons, vegetables, alfalfa, and cotton. Hamburg was a great innovator and was known for providing his farm workers with good housing and working conditions. In the 1950s he went to Israel as a consultant to the Ministry of Agriculture and commuted regularly between California and Israel. Hamburg died in 1976 at age 78. He was remembered as an outstanding farmer who dedicated his life to agricultural development. Goldsmith elaborates on Hamburg's life and his impact on people in the US and Israel who sought his advice and assistance. (See also: Gruber, Ruth. He Makes the Desert Bloom. *Reader's Digest* 71(424):107–122, August 1957; and Hamburg, Alice. *Sam Hamburg, Agricultural Pioneer in California and Israel.* Berkeley, CA, Western Jewish History Center, 1989.)

73. Greenwood, N. H. (1973). Sol Barth: A Jewish Settler on the Arizona Frontier. *Journal of Arizona History*, 14(4): 363–378.

Solomon Barth (ca. 1843–1928) came to the United States from East Prussia in 1856 and headed West. He was involved in many business ventures and eventually settled in Arizona where he made his fortunes in cattle, farming, trading, and gambling. In a card game with Mexican ranchers, he won cash, squatters' rights, water rights, and several head of sheep on the Little Colorado River at the juncture of two developing trade routes. He expanded his holdings, built a dam to divert water to his property and established the town of St. Johns which became a major trade route. Barth was involved in conflicts with Mormon traders, he became active in local politics, and served a two year prison term for corruption. In time, Barth reconciled with the Mormons and even requested that his funeral services be held in the Mormon Church. He survived Indian raids, gunfights and lawlessness to became a major player in the region. This is a fascinating biography of a Jewish entrepreneur who made a life in the wild west during the early days of westward expansion.

74. Gruber, Ruth (1957 August). He Makes the Desert Bloom. *Reader's Digest*, 71(424): 107–111.

A biographical profile of Sam Hamburg, who developed the California desert near Fresno into an abundant agricultural area. Gruber describes how Hamburg overcame disasters, floods, failures, and crop damage to shape an agricultural enterprise of 12,000 acres of cotton, cantaloupes, grain, seed crops, and produce with a market value of five million dollars and annual gross income exceeding two million dollars. Hamburg also served as an agricultural advisor to the government of Israel. He taught agricultural practices to Israeli farmers and is credited with bringing nearly 15,000 acres of former Israeli swampland under cultivation with cotton, sugar beets, peanuts, rice and vegetables and successfully bringing irrigation to a desolate area in the Beit Shean Valley which became an agricultural showplace. (See also: Goldsmith, Steven R. Sam Hamburg World's Foremost Jewish Farmer. *Western States Jewish Historical Quarterly*, 10[4]:330–342, 1978; and Hamburg, Alice. *Sam Hamburg, Agricultural Pioneer in California and Israel.* Berkeley, CA, Western Jewish History Center, 1989.)

75. Hallberg, Carl V. (1989). Jews in Wyoming. *Annals of Wyoming*, 61(1): 10–31.

This article offers a fascinating history of Jewish pioneers who came to Wyoming in the late nineteenth and early twentieth centuries seeking a life in the Western frontier. Economic developments like transportation, agriculture, and mining attracted Jewish immigrants to Wyoming from 1867–1910. The earliest representatives came in the 1860's to 1870's as soldiers, miners, sheep ranchers, and merchants. Independent Jewish homesteaders and ranchers settled in Laramie, Platte, Goshen and Carbon counties. Attempts to form agricultural colonies in began in 1900 in the Torrington and Wheatland districts with support from the Jewish Agricultural Society, and the Jewish Agriculturists' Aid Society (Chicago). Hallberg gives details of the settlement plans, land, soil and weather conditions, and the political and social factors that settlers faced. Economic and credit problems, foreclosures, declining farm prices, and the government's failure to undertake reclamation efforts that would have benefited the region, led to failure of the settlement experiment. By 1933 most of the farmers had abandoned their farms and moved to California, some remained in Wyoming and entered other professions.

76. Harris, Lloyd David (1981 February). Lest We Forget Beersheba. *Wichitan*, 3(10): 49–51, 60–63.

A history of the agricultural colony at Beersheba, Kansas. The colony was established in July, 1882 under the auspices of the Hebrew Union Agricultural Society (HUAS) by Rabbi Isaac Mayer Wise, editor of the *American Israelite.* Twenty six Russian Jewish families from Cincinnati settled in Beersheba, north of Cimarron and a few miles northeast of the town of Kalvesta. Although they were provided with housing and farm implements, the colony faced many difficulties. The settlers had no farming experience and the land was unsuitable for planting, it was treeless and lacked water. HUAS provided implements and supplies on a loan basis and placed restrictions on colonists' decision making on their own tracts. Some took off farm jobs to earn additional income. The colonists could not agree on how to operate the settlement and some began to leave; by 1890 all had left and the colony was abandoned. Some settlers remained in the area and became merchants in the nearby towns of Ravenna and Eminence. Includes photos of crumbling buildings, and the old Beersheba Ravenna cemetery.

77. Hornstein, Frank D. (1986 Summer). Sowing Anti-Semitism: The Jewish Stake in the Farm Crisis. *Reform Judaism*, 14(4): 22–23.

Report of a meeting held at Temple Israel in Minneapolis to discuss the farm crisis in the Midwest and the anti-Semitism in the region which blames the farmers' economic woes on Jewish bankers and the Jewish/Zionist conspiracy. The

"Farmer Urban Dialogue" sponsored by the Jewish Community Relations Council-ADL of Minnesota and the Family Farm Organizing Resource Center was attended by farmers, rabbis, Jewish communal leaders and representatives of progressive farm organizations, focused on strategies for combating the activities of various hate groups that are spreading anti-Jewish propaganda. The most effective way to counter the propaganda is to have more dialogues and face to face meeting between farmers and Jews and to seek an economic solution that will give farmers a fair return on their investment and enable them to remain in business.

78. Huhner, Leon (1941). Moses Elias Levy, an Early Florida Pioneer and the Father of Florida's First Senator. *Florida Historical Quarterly*, 20(4): 319–345.

Moses Elias Levy (1782–1854) was born in Morocco, brought up in Gibraltar and eventually migrated to St. Thomas (Virgin Islands) where he made his fortunes in the lumber business. In later years, he became interested in Florida where he believed there were opportunities for colonization and development. He acquired land in various parts of Florida and continued to expand his holdings. Levy was a devoted Jew who wanted to establish a colony as a refuge and religious settlement for the Jews in Europe who were facing oppression. He traveled widely throughout Europe and the United States trying to find support for the project. He established a sugar plantation called "Pilgrimage" at Micanopy in Alachua County. The first group of colonists arrived in 1822 from Germany and France and others arrived from New York and New Jersey. He built roads, erected houses, furnished provisions, stables, a cornhouse, a blacksmith shop and a supply of corn to last for a year. Levy was beset by litigation in connection with his land purchases and questions regarding his citizenship. The sugar house and its contents were set on fire by hostile Indians in 1835. The colony was abandoned and later destroyed to keep it from falling into the hands of Indians. Levy continued to be active in public life and the welfare of Florida. His fortunes began to dwindle and he died in White Sulphur Springs, Virginia in 1854. Huhner presents many details about Levy's life, his marriage, relationship with his four children, his opposition to slavery, and his work on interfaith relations among Christians and Jews.

79. Klauber, Laurence (1970). Abraham Klauber — A Pioneer Merchant (1831–1911). *Western States Jewish Historical Quarterly*, 2(2): 67–90.

Abraham Klauber lived in New Orleans, St. Louis, New York and moved to Volcano, California fifty miles east of Sacramento in 1852 and established a retail store that served miners, farmers and the public. He tried to establish businesses in other locations which included Genoa and Carson City, Nevada where he bought a 189 acre ranch in 1859 and in 1887 he acquired a 2400 acre stock and timber operation in Mendocino County. The ranches were managed by ranch operators and although they were not profitable ventures they were used by the family for summer vacations. (See also: Angel, Myrick. *History of Nevada, with Illustrations and Biographical Sketches of Its Prominent Men and Pioneers.* Oakland, CA, Thompson and West, 1881).

80. Kohanski, Dorothy (2003 Spring). Imagining My Painted Woods Great Grandfather. *Roots—Key: Newsletter of the Jewish Genealogical Society of Los Angeles*, 23(1): 23–25.

Using information gathered from a genealogical search of her roots, Kohanski holds an imaginary conversation with her grandfather, Solomon Dellar, who emigrated from Russia with his wife Rachel, three sons and two daughters in 1881 and joined the Painted Woods agricultural colony near Bismarck in the Dakota Territory. The Dellars remained at the colony five years and then moved by wagon train to Portland, Oregon. Kohanski, speaking for Dellar, discusses his life at Painted Woods, and why they decided to leave. In Portland, the sons became shoe and clothing merchants and Dellar sold farm equipment. Dellar died in the flu epidemic in 1918.

81. Kohler, Max J. (1914). An Early American Hebrew-Christian Agricultural Colony. *Publications of the American Jewish Historical Quarterly*, 22: 184–186.

Details of a plan by the American Society for Meliorating the Condition of the Jews to establish an agricultural colony for Jews who converted to Christianity. A farm was leased in Harrison, New York in Westchester County (31 miles from New York City). Agents were appointed to acquaint European Jews with the colony and thus secure their conversion and migration. The plan was deemed unworkable and abandoned within one year.

82. Kohler, Max J. (1900). Some Early American Zionist Projects. *Publications of the American Jewish Historical Society*, 8: 75–118.

This paper begins with a short history of Jewish settlement activities in Palestine and North and South America. The remainder of the paper focuses on Mordecai M. Noah, a diplomat, editor, playwright, lawyer, and judge who wanted to establish a place in America where Jews can settle and have their own government in a setting that will serve as a model for others. He proposed the establishment of the City of Ararat on Grand Island near Buffalo, New York as an ideal location for the colony. The paper presents Noah's vision for the colony, his perceptions of its importance for Jews and his efforts on behalf of the colony. Noah made contacts with European and American leaders attempting to gain support for the project. He held open meetings and made public statements and a held a celebration to declare the opening of the colony. However, the project was widely criticized and never realized. Kohler indicates that Noah never actually bought the land and never set foot on the Island. Includes excerpts of a paper presented before the Buffalo Historical Society describing the project and the text of Noah's "Proclamation to the Jews" setting forth his vision for the colony.

83. Kramer, William M., and N.B. Stern (1985). Isaac Lankershim of the San Fernando Valley: Jewish-Born Baptist. *Western States Jewish History*, 18(1): 67–75.

Biography of Isaac Lankershim, who was born Jewish in Bavaria in 1818 and migrated to the United States in 1836. He settled for a short time in Missouri where he began raising cattle. He married in 1842 and converted to the Baptist faith. In 1854 he migrated to California and began raising wheat in the Napa Valley and expanded to livestock and grain enterprises in Stanislaus, Merced, Fresno, San Diego, and Los Angeles counties and played an important role in the agricultural development of the San Fernando Valley. Lankershim was also a miller and major exporter of wheat to European markets. Although a convert, Lankershim maintained business ties with the Jewish communities of San Francisco and Los Angeles. He died in 1882. Lankershim was remembered as " a farmer and cattleman who became one of the most significant agricultural pioneers of California."

84. Krauskopf, Joseph (1913 July). The Lost Art Retaught. *American Citizen*, 3(1): 29–30–64.

The National Farm School, located in Bucks County, PA, 30 miles from Philadelphia, [today Delaware Valley College] trains Jewish students in practical and scientific agriculture. Krauskopf, Founder and President describes the school and its grounds and facilities and the importance for training Jewish youth who want to farm.

85. Lease, Richard J. (1977). Eugene J. Stern: Merchant, Farmer and Philanthropist of Las Cruces, New Mexico. *Western States Jewish Historical Quarterly*, 9(2): 161–166.

A profile of Eugene J. Stern, who emigrated to the United States from Hungary in 1903. He had jobs in retailing, insurance, and dry goods in Texas and Colorado and eventually moved to Las Cruces, New Mexico where he owned a general store. In 1928 Stern established a 1600 farm in the Rio Grande Valley that employed 50 farmhands. In 1954, he sold 800 acres and farmed the remaining 800 acres for the next ten years with net earnings about $40,000 annually. Stern was a philanthropist who supported civic, education, and religious activities in Las Cruces. He was a major benefactor in construction of the first synagogue in Las Cruces and was instrumental in establishing the first Dona Ana County Fair and the Chamber of Commerce.

86. Lemieux, Christina (1988). Michigan Utopia: The Sunrise Cooperative Farm Community. *Chronicle: The Bi-Monthly Magazine of the Historical Society of Michigan*, 23(6): 4–8.

In June 1933, a group of "Jewish libertarian anarchists" purchased 9000 acres in Saginaw County, Michigan for a collectivist agricultural community based on voluntary agreement and mutual cooperation. The colony, Sunrise Cooperative Farm Community, began operation on June 26, 1933 with 85 families comprising about 300 people. They constructed housing units, set up elementary and high schools and published a newspaper. However, the anarchist nature of the community made it difficult to enforce the rules since no person or group had any real power. The experiment failed and the farm was sold to the Resettlement Administration in December, 1936. Lemieux claims that Sunrise Community did not succeed for two major reasons, factionalism and lack of motivation. Factionalism and bickering was widespread because each person felt that his/her view should prevail and the lack of an organizational structure left the group powerless to enforce decisions. Many members were highly motivated and worked long hours. But a small group wanted free choice regarding the type and amount of work they perform. This group preferred to spend their time discussing political and social issues or prolonging a membership meeting for hours. Some called in sick frequently or took long vacations. This caused a rapid turnover in membership and destroyed the morale and motivation of the dedicated members. About 25 motivated families wanted to try the experiment again. They acquired a 640 acre farm in Samos, Middlesex County, Virginia in 1937. Despite dilapidated buildings, no plumbing or electricity, and poor roads, they were determined to make it a functional community. But they again encountered factionalism and lack of motivation and financial difficulties that plagued the Michigan colony. Members began to leave and the farm was abandoned when the last family left in April 1940.

87. Levin, I. Harry (1978). History of Alliance, New Jersey, First Jewish Agricultural Settlement in the United States. *Vineland Historical Magazine*, 54(1): 1–14.

The Alliance Agricultural Colony in Cumberland County, NJ was first settled in 1882 by 43 Jewish immigrant families from Russia with financial assistance from the Hebrew Immigrant Aid Society and the Jewish Agricultural and Industrial Aid Society. Although they lacked farming experience, adequate capital and had to contend with bad soil, they persevered and remained on the land mainly as poultry farmers. More immigrants arrived and the colony continued to grow. Some of the newer settlers established small factories which transformed Alliance into a mixed enterprise colony that helped it survive. Levin discusses the families who settled there, the synagogues and organizations they established, and the visitors and philanthropists that gave time and money to provide religious and secular education. Following World War II, many immigrants left Alliance and moved to the cities to enter other businesses. By the 1970s most of the original settlers were gone and only a handful were still farming. The settlers' children moved away seeking opportunities in non-farm occupations. Levin claims that the Alliance Colony produced a great number of persons who attained state and national prominence. He mentions some of those persons and their accomplishments.

88. Levinson, Robert E. (2002). Jews and Jewish Communities on the Great Plains, 1820–1977. *Western States Jewish History*, 35(1): 36–59.

The westward movement which began in the 1820's was attractive to Jewish immigrants from Eastern Europe. As word of the new opportunities in the West reached immigrants in the urban areas of the East, many made the trek west. They became peddlers, traders, merchants, farmers, politicians, and other professions. Immigrants who wanted to make their livelihoods as farmers settled in colonies in Kansas, North Dakota, South Dakota, and Colorado that were established with the aid of organizations like the Jewish Agricultural and Industrial Aid Society and the Hebrew Union Agricultural Society of Cincinnati. Most of the colonies were not successful and were abandoned after a few years. Levinson points out that lack of agricultural skills, lack of care and follow-up by sponsoring organizations, harsh winters, unsuitable land, and high interest rates on loans as reasons why the colonies did not succeed. Some individual farmers remained on the land and became successful farmers. For example, Sam Bober of Newell, South Dakota, a graduate of the Baron De Hirsch Agricultural School at Woodbine (NJ), was an organizer of the county Farm Bureau and a plant breeder who developed rust resistant strains of wheat, grasses, and seeds.

89. Levy, Abraham R. (1978). Central North Dakota's Jewish Farmers in 1903. *Western States Jewish Historical Quarterly*, 11(1): 3–17.

Rabbi Abraham Levy was a German immigrant who came to the United States in 1881. He served as a Rabbi in Erie, Pennsylvania; Athens, Georgia; Waco, Texas, and Chicago, Illinois. He was a leader of the "back to the soil" movement for Eastern European immigrants and founded the Jewish Agriculturists' Aid Society of America. The Society's purpose was to give loans to immigrants who wanted to buy land and enter farming. This is Levy's account of visits to 10 Jewish families living on 160 acre farms in North Dakota. He describes their farms, family life, and living and working conditions. His overall impression was favorable and he felt that the families he visited were satisfied with life on the farm. (Reprinted from Charles S. Bernheimer, editor. *The Russian Jew in the United States.* Philadelphia, John C. Winston, 1905, pp. 392–403. Edited, with an introduction and annotations by Norton B. Stern and William M. Kramer.)

90. Lifschutz, Ezekiel (1966 December). Jacob Gordin's Proposal to Establish an Agricultural Colony. *American Jewish Historical Quarterly,* 56(2): 151–162.

Jacob Gordin (1853–1909) was born in Russia, had a religious education and became a journalist and teacher in a Talmud Torah. He created a sect of his own called "Bratsvo" that rejected Talmud and post-biblical teachings. His sect accepted only the ethical teachings of the bible and denounced trade and commerce as professions for Jews. He felt that the most desirable occupation for Jews to pursue is agriculture. Gordin migrated to the United States in 1891 and contacted the Baron de Hirsch Fund to assist in finding a location to establish an agricultural colony based upon the Bratsvo philosophy. He toured the Woodbine colony as a possible location but found it unsuitable. Gordin abandoned the plan and directed his efforts to theater. He became a famous Yiddish playwright who wrote 100 plays during an 18 year period. (Also published in: Karp, Abraham, Editor. *The Jewish Experience in America. Volume IV—The Era of Immigration.* New York, Ktav, pp. 253–261, 1969.)

91. Loeb, Morris (1911 December 23). The Jewish Settlers. *Survey,* 27(27): 1403–1404.

Loeb discusses issues raised at the third convention of the Federation of Jewish Farmers. The Federation is an organization of Jewish farmers involved in self help programs to enable them to make a living in agriculture. Loeb cites five factors that influence Jews from Eastern Europe to choose farming as a lifestyle—love of the land, previous experience in agriculture, individual motivation, a healthy work place, and a desire to remove children from urban environments. Loeb is critical of the motives of the Jewish Agricultural and Industrial Aid Society which he claims is settling immigrants on farms in order to disperse them to rural areas. However, the Federation of Jewish farmers has been successful in strengthening the farm families and helping them to remove obstacles that enable them to remain on the land. Topics of interest at the convention included access to good schools, local cooperation, credit unions, agricultural banks, rural schools, cooperation for purchase and sale, sanitation, protection of property and person, and satisfactory means of travel and communication.

92. Marcosson, Isaac Frederick (1914 March 21). Rural Credit Unions and Cooperative Farming: Where Dollars Are Neighborly. *Colliers*, 53(1): 9-10, 17, 31–33.

A look at Jewish Farmers Credit Unions and their benefits. Marcosson visited a Jewish farm settlement in Fairfield, Connecticut to observe the operation of its credit union. The credit union offers loans, mortgages, central purchasing, cooperative milk handling, mutual insurance and other services. He describes some Jewish farmers credit unions in Ulster and Sullivan counties (New York) that offer cooperative fire insurance and a cooperative creamery. He praises the cooperative idea as a collective effort applied to a multitude of needs.

93. Marcus, Jacob Rader (1956). Trail Blazers of the Trans-Mississippi West. *American Jewish Archives*, 8(2): 59–130.

From the 1830s through the 1880s, there was a large influx of Jewish immigrants from Germany and Western Poland to the United States. Julius Stern, editor of a German Jewish Weekly newspaper and Rabbi Isaac Lesser publisher of *The Occident* and *American Jewish Advocate* and Rabbi Isaac Meyer Wise founder of *The American Israelite* encouraged the new immigrants to settle in the territories west of the Mississippi River. Those who heeded the call moved to cities and rural areas in Missouri, Arkansas, Texas, Iowa, New Mexico, Colorado, Kansas, Arizona, Utah, Minnesota, Nevada, The Dakotas, Montana, Nebraska, Oklahoma, Wyoming and Idaho. They entered a variety of professions, established businesses and became active in Jewish and secular activities in their communities. Some of these pioneers became farmers, cattle ranchers, and cowboys; others became merchants who and sold and traded agricultural machinery, equipment, and products. This article offers glimpses of life in those communities gleaned from newspapers, memoirs, letters, synagogue archives, tombstones, and other records.

94. Marschall, John P. (2002). Jewish Agricultural Experiment, Wellington, Nevada, 1896–1902. *Agricultural History*, 76(2): 244–260.

Describes the history and philosophy that led to the establishment of a Jewish agricultural colony in Wellington, Nevada in 1896. Joseph Nudelman, a farmer and leader of the Wellington initiative, persuaded about 30 immigrant families from Eastern Europe to settle on the 5500 acre colony. Marschall describes Nudelman's ideology and the financial arrangements to get the settlement started. The colony was disbanded in 1902 for a variety of reasons that included poor soil, lack of water, disagreements among settlers, mismanagement, and corruption and theft. Nudelman left Nevada and settled in Portland, Oregon; some families remained in the area and continued farming.

95. Mauk, Cathy (1994). Fargo, North Dakota's Lonely Jewish Cemetery. *Western States Jewish History*, 26(3): 251–255.

Description of a cemetery in North Dakota, 20 miles north of Devil's Lake near the village of Garske, containing about two dozen graves dating from the late 1800s to the turn of the century. The cemetery served as a resting place for Jewish farm families who lived in the agricultural settlement near Iola called the "Garske Colony." Mauk gives details about the short lived colony and anecdotes about some of the people buried in the cemetery.

96. Menes, Abraham (1949). The Am Oylom Movement. *YIVO Annual of Jewish Social Science*, 4: 9–33.

The Am Oylom [also called Am Olam] movement was founded in Russia in the 1870s to fill the need for Jewish emancipation through emigration to a place where they can establish an independent territory. Members planned to emigrate to America and establish socialist oriented Jewish colonies based on agriculture. Menes analyzes Am Oylom's philosophy, the views of its founders and the attempts to establish agricultural settlements in Louisiana, South Dakota, and Oregon. The colonies faced numerous previously unanticipated difficulties which led to their failures in a few short years. Despite early indications of success they were confronted with crop failures, poor soil conditions, lack of agricultural training, and illness. Conflicts arose regarding economic, psychological and political differences, and isolation from other Jews. When the colonies disbanded, the settlers moved to urban areas and entered other professions and the movement came to an end. (Also published in: Fishman, Joshua A. *Studies in Modern Jewish Social History [1972]*, New York: Ktav, 1972, pp. 155–179, and Gurock, Jeffrey S. Editor, *American Jewish History, Volume 3. East European Jews in America, 1880–1920 Immigration and Adaptation.* New York: Routledge, 1998, pp. 109–133.)

97. Monaco, Chris (2002). A Sugar Utopia on the Florida Frontier: Moses Elias Levy's Pilgrimage Plantation. *Southern Jewish History*, 5(1–5): 103–140.

A study of the activities of Moses Elias Levy (1782–1854) a pioneer plantation owner and anti-slavery activist. Focuses on Levy's goal to establish a plantation in Alachua County, Florida to serve as a settlement for Jewish refugees from Eastern Europe. Presents Levy's philosophy regarding redemption of oppressed Jews of Europe and the establishment of the Pilgrimage Plantation near Micanopy, Florida. Describes the plantation, living arrangements, dwellings and backgrounds of some of the settlers and life on the plantation. Despite attempts to attract settlers, only about five Jewish families and a number of slave families settled at Pilgrimage. The low number of settlers, criticism from other Jewish leaders, legal challenges to his land ownership, and the realization that the settlement idea had failed, led to the abandonment of Pilgrimage in 1825. Levy moved on to other social causes such as battling anti-Semitism, Jewish-Christian relations, universal education, and abolition of slavery. (For a definitive biography on Moses Levy see: Monaco, Chris S. *Moses Levy of Florida: Jewish Utopian and Antebellum Reformer*. Baton Rouge, LA, Louisiana State University, 2005.)

98. Muehl, Siegmar (1999). Isidor Bush and the Bushberg Vineyards of Jefferson County. *Missouri Historical Review*, 94(1): 42–58.

A study of the life of Isidor Bush (1822–1898) who came to the United States from Vienna in 1866 and settled in St. Louis. He became a businessman, publisher, and civic leader. Bush took an interest in grape cultivation and began to pursue his career as a vintner by purchasing 241 acres in Jefferson County, Missouri in 1865 for to cultivate established varieties and develop new strains. His foreman, Gustave Meissner, a German immigrant and an expert in grape propagation, was hired as foreman and eventually became Bush's partner. The company issued a descriptive, encyclopedic catalog and began to sell vineyard products worldwide. The catalogs made the company quite famous and it was known as a place to get vine cuttings for selective grape breeding. In the 1870s, Bush cuttings were used to save the French wine industry during a disease infestation. The business declined in the late 1880s due to competition with the California wine industry and went out of business in the early 1890s. Bush and Meissner died in 1898 and their sons dismantled the company.

99. Naftaly, Phillip R. (1991). Jewish Chicken Farmers in Petaluma, California 1904–1975. *Western States Jewish History*, 23(3): 231–247.

For three decades, Petaluma, California was known as the "Egg Basket of the World." This thriving community of chicken farmers began when Sam Melnick, an immigrant from Lithuania, came to the area in 1904 and established a seven acre ranch with 500 pullets. Many Jewish settlers interested in an agricultural lifestyle followed and by 1917 there were 25 families. Growth continued through the 1940s when there were between 200 and 300 farm families in the Petaluma area with two thirds engaged exclusively in raising broilers and fryers. The community began to diminish in the 1950s with the dwindling number of small scale farms and by 1975 less than ten families remained in the poultry industry. Naftaly describes the community, some of the farm families, social life, religious activities, and political differences among the settlers.

100. Narell, Irena Penzik (1975). Bernhard Marks: Retailer, Miner, Educator, and Land Developer. *Western States Jewish Historical Quarterly*, 8(1): 26–38.

Bernhard Marks (1832–1913) was born in Poland and arrived in the US at age 3 with his parents who settled in Providence, Rhode Island. He traveled to California in 1892 and worked at various jobs including the operation of several mines but met with little success. He married in 1859, and managed a grammar school in San Francisco. In 1874, he bought land in a farming colony near Fresno and raised livestock and raisins. Marks experimented with irrigation but found the soil unsuitable. When he failed at farming he got involved in land deals and real estate in Sacramento and Fresno counties. Marks died in 1913 and is remembered as a participant in Fresno's early development efforts.

101. Ornstein-Galicia, Jacob (1989). An American Jewish Family's Farm Odyssey. *American Jewish Archives*, 41(1): 53–76.

A professor of Modern Languages and Linguistics at the University of Texas, El Paso, offers a nostalgic look back at his youth on a farm in Thompson, Ohio about 45 miles northeast of Cleveland. He explains how his immigrant parents got involved in farming and became the only Jewish farm family in Thompson and one of the few in Geauga County. Ornstein-Galicia reminisces about the hardships and pleasures of farm life and how his parents coped with the daily routines and relationships with the community. His narrative offers memories about the farm, work, animals, school, and the contributions of Jews to American agriculture.

102. Osofsky, Gilbert (1960 March). The Hebrew Emigrant Aid Society of the United States (1881–1883). *Publications of the American Jewish Historical Society*, 49(3): 173–187.

The Hebrew Emigrant Aid Society was established in 1881 to support Jewish refugees from Russia who were fleeing persecution. Although it lasted only two years, and was disbanded in 1883, the Society supported and helped settle about 14,000 immigrants. It was involved in dispersing immigrants to agricultural colonies and provided loans and economic assistance to settle farmers in Louisiana, Colorado, Kansas, South Carolina, Texas, Ohio, Oregon, and the Dakota and Washington territories. This article is a mix of material in contemporary newspapers of the time about the society and its philanthropic and settlement activities. (Also published in: Karp, Abraham J. Editor. *The Jewish Experience in America. Volume IV—The Era of Immigration.* Waltham, MA., American Jewish Historical Society, 1969, pp. 73–85.)

103. Palitz, Bernard A. (1909 July). American Jewish Farmers. *Jewish Comment*, 29(14): 219–220.

In this address presented at a meeting of Jewish Social Workers in Buffalo,

NY, Palitz makes a case for encouraging Jewish immigrants to choose farming as an occupation. He takes issue with those who say that the Jew is not suited for life on the land. Agriculture offers immigrants the opportunity to leave the crowded cities and tenements, and the unhealthy atmosphere of factories and pursue a life on the land where opportunity awaits. Palitz comments that farming will enable immigrants to become self sufficient and live in a healthy environment. He urges Jewish social workers to support the establishment of a national organization to promote agriculture among Jews.

104. Papermaster, Isadore (1978). A History of North Dakota Jewry and Their Pioneer Rabbi. *Western States Jewish Historical Quarterly*, 10(1): 74–89, 10(2):170–84. 10(3):266–93.

A three part history and biography (by his son) of Rabbi Benjamin Papermaster, the first Rabbi to serve in North Dakota. Rabbi Papermaster was the only Rabbi serving Jews in all of North Dakota and western Minnesota until the turn of the century. He settled in Grand Forks in 1890 where he served as spiritual leader and teacher for the Jewish settlers in the regions who were mainly farmers and shopkeepers. Part I covers aspects of Papermaster's life in Kovno, Lithuania, his immigration to America and his decision to go to North Dakota. Part II gives accounts of his service as rabbi, *schochet*, teacher, consultant, and mentor for the area's Jewish community. Part III covers the early twentieth century when Jewish settlement in North Dakota was increasing in the towns along the route of the Great Northern railway. Under Papermaster's leadership, the Jewish Community established a synagogue, Hebrew school, a Ladies Aid Society, a kosher meat slaughtering facility, and a burial society. His efforts enabled the farm families to retain their Jewish identity and keep their traditions and customs alive. He was active in politics and represented the Jews in local matters. Rabbi Papermaster died September 1934.

105. Parsons, Beatrice (1956 October). Starting from Scratch. *National Grange Monthly*, 53(10): 26–28.

A summary of the highlights of the Annual Report of the Jewish Agricultural Society. Indicates that 2000 of the "displaced persons and their families" who have migrated to the United States since the end of WWII, have been successfully settled on farms with the Society's economic and technical assistance. Some of the settlers joined cooperatives, constructed synagogues and educational centers and joined the Grange. Presents profiles of two outstanding Jewish farmers active in Grange. Louis Goldschmidt of Windsor, NY, member of the Susquehanna Grange, considered among the outstanding farmers of New York State and Siegfried Levy of Abbington, Connecticut who manages one of the highest producing dairy herds in Windham County.

106. Pearlstine, S. S. (1980). Jewish Homesteaders in Wyoming—1913. *Western States Jewish Historical Quarterly*, 12(1): 31–32.

Pearlstine claims that there were about five million acres available for homesteaders in Wyoming in 1913. He urges the organizations that are supporting Jewish settlement in agricultural areas to give consideration to the opportunities in Wyoming. He offers as examples successful farmers such as Abe Shapiro, a fruit farmer, Nathan Cohen, a former junk peddler who acquired homestead land, and the Massion Brothers who run a large farmstead.

107. Pincus, Joseph W. (1913 June). Jewish Problems in Rural Communities. *Jewish Charities*, 3(11): 11–13.

Pincus states that the settlement of Jewish immigrants in farming communities creates a need to improve social and educational facilities in rural areas. He suggests that Jewish social work among the settler families should include provisions for religious education and observance, organizations for the wives of farmers, country life organizations for farm youth, improvement of general educational facilities, access to libraries and books, and lectures and night classes for farmers.

108. Pincus, Joseph W. (1913). The Jew, a Successful Farmer. *American Citizen*, 3(2): 87–88, 109–110.

Pincus, editor of *The Jewish Farmer*, responds to critics who question whether Jews should pursue a life on the farm. He points to the efforts of the Baron De Hirsch Fund, and the Jewish Agricultural and Industrial Aid Society that have provided support to Jews entering agricultural settlements. There are also independent successful Jewish farmers who acquired their own farmsteads. Pincus indicates that there are about 5000 Jewish farmers in the United States with property holdings estimated at about $33 million. He points out that there is a natural longing among Jews to return to agriculture as evidenced by the increasing enrollment in the National Farm School and an increasing number of young Jews working as farm laborers to gain farm experience. When judging the suitability of Jews for careers in farming, one should look at the improving economic conditions that have enabled farmers to pay off their loans and the social, religious, and educational facilities available in their communities.

109. Pincus, Joseph W. (1911 July). Social Work among Jewish Farmers. *Jewish Charities*, 1(12): 8–11.

In this paper, read at the Conference of Jewish Social Workers, June 9, 1911, Pincus claims that the lack of social life on the farms is one of the main reasons why more Jewish immigrants do not enter farming. The publication of *The Jewish Farmer* and establishment of the Federation of Jewish Farmers provided information about other Jewish farmers and offered a means of social contact. To enhance education and social contact among Jewish farmers, Pincus recommends additional activities that include: agricultural libraries and farm demonstrations, lectures, citizenship classes, improvement of educational facilities, religious education, social amusements, circulating libraries, formation of children's and women's clubs. He offers recommendations on how to organize these activities and make them work well.

110. Pincus, Joseph W. (1903). A Significant Experiment with the Jews in Agriculture. *Independent* 55(2861): 2337–2343.

Pincus, a member of the Woodbine (NJ) agricultural settlement and Superintendent and Assistant Principal of the Baron de Hirsch Agricultural and Industrial school, offers his observations of the progress made by Woodbine from its early days to the present. It began with about 12 families on 5300 acres in Cape May County in Southern New Jersey purchased with funds from the Jewish Agricultural Society. Woodbine has developed into a center of trade, manufacturing, and education with a population of 2000 people. In addition to the production of fruits, vegetables, field crops, milk, eggs, and poultry, the settlement employs 400 people in factories that produce clothing, machinery, building materials, and knit goods. The Agricultural and Industrial School provides a quality education in agriculture and a broad curriculum of general studies. Pincus points out that graduates of the school have taken positions in New Jersey including the College of Agriculture at Rutgers University and other colleges and institutions throughout the country.

111. Plaut, W. Gunther (1965). Jewish Colonies at Painted Woods and Devil's Lake. *North Dakota History*, 32(1): 59–70.

Study of two agricultural colonies settled by Russian immigrants in Dakota Territory in the 1880s. The first, Painted Woods, was established in 1881 through the efforts of Rabbi Judah Wechsler, of St. Paul, Minnesota. Plaut describes Wechsler's activities on behalf of Painted Woods, justifications for the project, appeals to the Jews of St. Paul for assistance, and his belief that it was a project that would succeed. Painted Woods began with over 50 families owning 160 acres each. The colony lasted five years until 1886 when settlers left due to a number of problems including crop failures, lack of farming experience, poor soil, drought, hailstorms and prairie fires, debts, and internal feuds. A revival occurred in 1906 when 42 individual homesteads were established in the region and by 1912, 250 families were settled on about 5000 acres. The second settlement, at Devils Lake was established in 1882 with the help of the Baron De Hirsch Fund and support from Jews in Minneapolis. It began with 22 families who were joined by former Painted Woods members and members from a nearby colony at Iola.

Painted Woods was successful in its early years but began to dwindle in later years when disputes about funding, high interest rates and abject conditions emerged. The Devils Lake Colony was out of existence by the mid-1920s. Plaut explains the funding controversies, rivalries among St. Paul and Minneapolis supporters and the conditions of life on the settlements that led to their demise.

112. Price, George M. (1958). The Russian Jews in America. Part I. (Translated from the Yiddish by Leo Shpall.) *Publications of the American Jewish Historical Society*, 48(1): 28–62.

This article which originally appeared in the Russian press describes the lives of Russian immigrants in America in the decade between 1881 and 1891. Price examines the reason for leaving Russia, the number and types of immigrants, attitudes of American Jews toward the newcomers, living conditions, employment and economic status, assistance from philanthropic societies and the impact on world Jewry. Includes agricultural work and the Baron De Hirsch fund which is presented in greater detail in part 2. (Also published in: Karp, Abraham J. editor. *The Jewish Experience in America. Volume IV—The Era of Immigration.* Waltham, MA., American Jewish Historical Society, 1969, pp. 265–299. See also: Price, George M. The Russian Jews in America. Part 2. Jewish Agricultural Colonies in America. *Publications of the American Jewish Historical Society*, 48[2]:78–133, 1958.)

113. Price, George M. (1958). The Russian Jews in America. Part 2. Jewish Agricultural Colonies in America. (Translated from the Yiddish by Leo Shpall.) *Publications of the American Jewish Historical Society*, 48(2): 78–133.

Price analyzes the attempts by Russian immigrants to establish agricultural colonies in Louisiana, South, Dakota, Kansas, Oregon and three colonies in New Jersey—Alliance, Carmel, and Rosenhayn in the 1880s. Some immigrants were attracted by the desire for a better life in America and the promises by the Baron De Hirsch Fund to support the colonization efforts which gave them hope of a bright future. Most of the colonies lasted only a short time and proved to be unsuitable for the new immigrants. Price attributes this to a number of factors including shortage of funds, poor land, unfavorable climates and lack of agricultural training. The New Jersey colonies survived because they were better financed, nearer to cities, and were designed as mixed agricultural/industrial enterprises that generated steady income despite continual mortgage and debt problems. The Baron De Hirsch Fund was a significant force in providing funding and support for the immigrants and helping them get acclimated to American life. (Also published in: Karp, Abraham J. Editor. *The Jewish Experience in America. Selected Studies from the Publications of the American Jewish Historical*

Society. Waltham, MA, American Jewish Historical Society, 1969. Volume IV, pp. 300–355. See also: Price, George M. The Russian Jews in America. Part 1. *Publications of the American Jewish Historical Society* 58[1]:28–62, 1958.)

114. Ragins, Sanford (1965 November). The Image of America in Two East European Hebrew Periodicals. *American Jewish Archives*, 17(2): 143–161.

During the mass immigration of Russian and Polish Jews in the 1880's, the East European Hebrew press began publishing articles on life in the United States. This article summarizes the coverage by *Hamelitz* published in St. Petersburg (Russia) in 1983 and *Hatzfirah*, published in Warsaw in 1884. The articles covered American Jews, their religious practices, social and charitable activities and how commerce, freedom, and employment affected them. There was great interest in the agricultural communities founded by Russian and Polish immigrants throughout the United States. These newspapers reported on the colonies in North Dakota, New Jersey, Louisiana, Arkansas, and Colorado.

115. Raider, Mark A. (1995). Jewish Immigrant Farmers in the Connecticut River Valley: The Rockville Settlement. *American Jewish Archives*, 47(2): 213–242.

The Rockville Settlement, established in 1900, was one of the most successful Jewish agricultural settlements in the Northeast. Yiddish speaking Eastern European Jewish immigrants with a love of the land and a commitment to a communal life, established a settlement in the Rockville-Vernon-Ellington area of the Connecticut River Valley. Although they represented a range a different religious traditions and political perspectives, the settlers were united by a spirit of community and a desire to make a success in farming. There were 72 members by 1909 and as late as 1953, 90 percent of the membership were still farming. Raider presents a detailed early history of the settlement, its farming practices, communal life, philosophy, reasons for its success, and its influence in the region. Raider was critical of the report of the Dillingham Commission, a congressional committee investigating the status of immigration to the United States. The report contained many negative stereotypes of Jews but indicated that the success of the Rockville settlement was due to "exceptionally intelligent Hebrews" who were the "best type" of Hebrew farmers. (Also available online at: http://www. ellingtonshul.org/ourcommunityshistory.htm.)

116. Reis, J. C. (1907). History of the Alliance Colony. *Menorah*, 42(4): 167–173.

The Alliance Colony was established on May 1882 in Salem County, New Jersey five miles from Vineland by twenty five Russian immigrant families who were mainly small traders and storekeepers. With assistance from the Hebrew Emigrant Aid Society (HEAS) they built and equipped houses, cleared land, and began planting vegetables. They earned income selling their produce and doing part time work for others as farmers and tailors. Reis explains how the settlers lived and worked and how they managed to overcome the lingering debts. The Baron De Hirsch Fund assisted with loans and in 1900 Alliance built a canning factory to can and market their produce. The factory generated additional income and improved the material condition of the settlers.

117. Rikoon, J. Sanford (1998). The Jewish Agriculturalists' Aid Society of America: Philanthropy, Ethnicity, and Agriculture in the Heartland. *Agricultural History*, 72(1): 1–32.

The Jewish Agriculturalists' Aid Society of America (JAAS) was instrumental in supporting several hundred Russian and Eastern European Immigrants who settled on farms in the Midwest from 1880 through 1930. The Society, with headquarters in Chicago, provided loans, and other forms of financial assistance to help the farmers survive and succeed. In addition to providing financial aid to the farm families, the Society undertook its own fund raising efforts to maintain its financial base. In 1900, it established a partnership with Jewish Agricultural and Industrial Aid Society (JAIAS), later the Jewish Agricultural Society of New York, which led to conflicts among the two groups. Lack of farming experience, concern about inability to maintain patterns of Jewish life, and lack of access to schools, led many farmers to leave for urban areas. The Chicago Society ran out of funds by 1908 and closed. The JAIAS in New York continued to support farmers who remained on the land. Rikoon explains the role of JAIAS as a force for Jewish farm settlement and the successes and failures through its years of existence.

118. Rikoon, J. Sanford (1997). An Overview of Jewish Farm Settlement in the Heartland of America. *Western States Jewish History*, 29(3): 117–146.

Chronicles the efforts of Jewish immigrants from Russia and Eastern Europe to establish agricultural communities in the Heartland from 1880 to 1930. Focusing mainly on the settlers in North Dakota, Rikoon points to three major initiatives that sparked the settlement efforts. 1. Individual families and small groups settling on government and railroad land between 1880 and 1910. 2. From about 1900 through the Depression era settlers preferred to purchase working farms close to urban areas with strong Jewish congregations. 3. The communal endeavors established as agricultural colonies between 1880 and 1940 which constituted less than 10 percent of the overall Jewish population on farms. Support for these efforts came from the Am Olam movement, the Jewish Agriculturalists Aid Society of America, and the Baron de Hirsch Fund. Rikoon discusses how the three settlement eras began, their supporters and detractors, and the economic, social, and cultural issues they faced. The majority of Jewish farmers did not stay on their original farms more than seven years and by 1965 most of the Jewish farmers were gone from the Heartland. Rikoon attributes this to several factors that include lack of farm experience, poor choices of land which was not productive, and low capital resources. Jewish families left farming because they were unable to keep the farms going for very long, they felt isolated from other Jews, Jewish social support networks, and educational facilities, and a desire to seek other opportunities and locations in order to survive.

119. Ringel, Mosheh H. (1939). Further Information Concerning the Colony "Sholam" on Yageville, Hill, Ulster County, New York. *Publications of the American Jewish Historical Society*, 35: 306–307.

Brief description of the Sholam Colony in the township of Warwarsing, near Rochester, New York. Ringel visited the site in 1936 and found crumbling buildings and ruins from the early colony. He points out that Jewish settlement in the area can be traced back to December 1837 when deeds were issued to 11 Jews who bought tracts of land. Ringel describes some of the early inhabitants, the terrain, various buildings (including the synagogue and mikvah) and historic grave markings. Although the colony lasted about four years, there is evidence that Jews settled in the area as early as 1818.

120. Roberts, Dorothy (1941 July). The Colony at Cotopaxi, Colorado. *Colorado Magazine*, 18(4): 124–131.

The Jewish agricultural colony at Cotopaxi, Colorado was established in 1882 through the efforts of Emanuel H. Saltiel, owner of a silver mine in Cotopaxi. He presented his plan before the Hebrew Emigrant Aid Society in New York and they offered about $10,000 to assist with settlement. Sixteen families (63 people) traveled to the site in spring and summer 1882 with only clothing and meager possessions. Roberts alleges that Saltiel failed to honor his commitments to provide housing, furniture, cooking utensils and the appropriate farm equipment. Settlers had to endure land that was not suitable for farming, bad weather, crop failure, and poorly constructed dwellings. The settlers complained to the Denver Jewish Community which dispatched members to investigate conditions at Cotopaxi. They prepared a report decrying the "destitute and pitiful condition of the colonists" and Saltiel's unsavory past. The Report recommended that HIAS abandon the project. Settlers were given $100 and ordered to leave for any destination they desired and the colony was out of business by 1884. Some settled in Denver, others went to Utah, California, and South Dakota. A list of the names, occupations, and native countries of the original settlers are given.

121. Robinson, Leonard G. (1912). Agricultural Activities of the Jews in America. *American Jewish Year Book*, 14: 1–115.

A detailed history of Jewish agricultural activities in the United States from 1820 to 1912. Robinson sets the stage with a description of agricultural colonies in Palestine, the Orient, Western Europe, Argentina, Brazil, and Canada. He then presents the American Jewish agricultural experience which offered new immigrants the opportunity to settle on communal farms around the country as an alternative to life in urban areas. He mentions colonies in California, Connecticut, Kansas, Louisiana, Michigan, Nebraska, New Jersey, New York, North Dakota, Oregon, Pennsylvania, South Dakota, Utah, Washington, Wisconsin, and Wyoming. Robinson analyzes the philanthropy of Baron de Hirsch who spearheaded the settlement efforts through the Jewish Agricultural and Industrial Aid Society. Agricultural schools and credit unions were established to help farmers obtain loans and education for their children. To illustrate the tendency of the children who grew up on these settlements to remain in agriculture, Robinson names over 12 individuals who continued to serve agriculture as professors, deans, entomologists, soil chemists, and farm school directors. In an "afterword" Morris Loeb outlines the factors that must be in place for Jewish farmers to succeed and remain in agriculture. These are financial support, agricultural education, instruction of the young, inducements to help them farm better, and communities that offer social, hygienic, and intellectual advantages for the residents. (Also published as: *Agricultural Activities of the Jews in America.* New York, American Jewish Committee, 1912. 96 pp.)

122. Rudd, Hynda (1981). The Mountain West as a Jewish Frontier. *Western States Jewish Historical Quarterly*, 13(3): 241–256.

A study of pioneer Jews who settled in the territories and states of Arizona, Idaho, Colorado, Montana, Nevada, New Mexico, Utah, and Wyoming. Rudd describes settlers who came west and made their livelihoods in a variety of trades and professions including agriculture. She presents brief profiles of cattlemen in Wyoming and Jewish agricultural colonies in Colorado, Utah, and California.

123. Rudin, A. James (1972). Bad Axe, Michigan: An Experiment in Jewish Agricultural Settlement. *Michigan History*, 56(2): 119–130.

In 1891, sixteen Russian Jewish immigrants purchased 640 acres to establish an agricultural colony near Bad Axe, Michigan, 125 miles north of Detroit and 50 miles east of Bay City. They called the colony "Palestine" and soon gained support from the Baron De Hirsch Fund and the Reform Jewish Community of Detroit, comprised largely of German Jews. Settlers planted crops, cleared land, and constructed buildings. At its height, the colony owned over 800 acres and had a population of approximately seventy people. They tried to make the experiment work but suffered a series of setbacks and settlers began leaving. The last family left in 1903. Lack of agricultural knowledge, poor financial support, defaults on payments, geographical isolation, poor soil, bitter weather, natural disasters, and a longing for better opportunities in cities were responsible for the colony's failure. It was further exacerbated by the ambivalence of the highly urbanized German Jews in Detroit to continue economic support of the Russian immigrants who were mainly Orthodox. Rudin points out that despite its setback, the Palestine Colony served as an impetus for formation of the Jewish Agricultural Society in 1900 and other settlement programs elsewhere.

124. Rudin, A. James (1968). Beersheba, Kansas: "God's Pure Air on Government Land." Kansas Historical Quarterly, 34(3): 282–298.

On July 26, 1882, twenty four Russian Jewish immigrant families comprising about sixty people, departed from Cincinnati to establish an agricultural settlement in Kansas called "Beersheba" located about 22 miles northeast of Cimarron in Hodgeman County. The settlement was encouraged and supported by the Hebrew Union Agricultural Society (HUAS) under the supervision of Rabbi Leo Wise, editor of *American Israelite.* This article discusses the rationale for the settlements, individuals involved in its formation and the problems encountered en route. The colony was disbanded four years later due to a variety of factors that included poor land, lack of agricultural experience, disagreement about HUAS management policies, weak financial support, isolation from markets, poor soil, harsh weather, and natural disasters. Many of the settlers had other professions or trades which they returned to after leaving the colony.

125. Ruxin, Robert H. (1977). The Jewish Farmer and the Small-Town Jewish Community: Schoharie County, New York. *American Jewish Archives*, 29(1): 3–21.

Presents case studies of several Jewish farm families in Schoharie County, New York from the end of WWII through the 1950's. With data gathered from oral interviews, Ruxin evaluates the demographic, socioeconomic, and cultural factors of the community and relationships between Jews and their Gentile neighbors. Includes an analysis of the development of the Schoharie County Jewish Community Center and the social forces that led to its breakup and disintegration.

126. Sabsovich, H. L. New Jersey's Colony of Russian Jews. *New Jersey Review of Charities*, 1(2): 29–32.

A brief description of the characteristics of the Woodbine colony consisting of sixty five individual farm units. Describes the rationale and aims of the colony and its importance as a manufacturing, agricultural, and educational center of Cape May, County. Includes facts about the industries (tailor, knitting, brickyards, and machine shops), housing facilities, home ownership, basic principles regarding work, religious and secular education, arrangement of the farms, and social life at Woodbine.

127. Saltiel, Miles (2005 Winter/Spring). The Jewish Agricultural Colony at Cotopaxi, Colorado: Rebalancing the Record. Part I. *Rocky Mountain Jewish Historical Notes*, 19(3,4): 1–7.

In May 1882, thirteen Russian immigrant families settled in an agricultural colony in Cotopaxi, Colorado. Others followed and the colony grew to twenty two families with a population of sixty three people. Due to a variety of social and economic factors, the colony lasted only two years and was disbanded in 1884. One of the major studies of Cotopaxi was a 1950 master's thesis *The Cotopaxi Colony* written by Flora Satt, a student at the University of Colorado. Satt attributed the failure of Cotopaxi to Emanuel Saltiel, a prosperous mine owner and businessman who procured the land for the settlement. Satt attributes the economic problems to Saltiel's mistreatment of the settlers and a failure to meet his commitments to assist them with housing, equipment, and supplies. Miles Saltiel, a British born relative of Emanuel Saltiel, traveled in the United States to do research on his family line. He became aware of the Satt thesis and declared that most of her contentions are unsupported and misrepresent the relationship between Emanuel Saltiel and the settlers. He claims that although his relative may have been overly optimistic in support of the Cotopaxi settlement, he believed it was a good enterprise for the area that would lead to economic benefits. Emanuel was not given enough information by the Hebrew Emigrant Aid Society (HEAS) which provided initial financial support for the colony and was misled into believing that the colonists would be able to meet their economic contracts on the land and housing that he offered. When it was clear that the colony was suffering from economic difficulties some of the settlers sought work at Saltiel's mine. They were not coerced into working in the mine as Satt alleges. The author claims that the colony's leaders were business people and proprietors who made good faith agreements with HEAS and Saltiel but the failure was due to their inability to meet the terms of the agreement. (See: Satt, Flora Jane. *The Cotopaxi Colony.* M.A. Thesis, University of Colorado [Boulder, CO], 1950. 83 pp.)

128. Saltiel, Miles (2005 Fall/Winter). The Jewish Agricultural Colony at Cotopaxi, Colorado: Rebalancing the Record. Part II. *Rocky Mountain Jewish Historical Notes*, 20(1,2): 1–11.

In Part II of the rebuttal, Miles Saltiel responds to the claims of bitterness and ill will between the colonists and Emanuel Saltiel as reported in Satt's thesis, translated accounts from the Russian press published after Satt's thesis, and a related article in the *Denver Tribune*. He focuses on issues regarding store credit, culture clashes, conspiracy theories, harsh weather, and distorted stories and errors. According to Saltiel, the colonists were aware of the debts they were accumulating but failed to pay them. Colonists who came from Ukraine knew how to function in the cold winters and had no difficulty withstanding the climate. Some settlers were earning money from various jobs in the community and were not destitute. Rounds of negotiations took place among the settlers, HEAS and Saltiel to deal with the growing debts. Saltiel attributes the colony's failure to unsuitability of the location for farming, a lack of working capital in the colony's budget, harsh pressures imposed by HEAS, the inexperience of the colonists, and the changes taking place in the agricultural sector.

129. Sapinsley, Elbert L. (1971). Jewish Agricultural Colonies in the West: The Kansas Example. *Western States Jewish Historical Quarterly*, 3(3): 157–169.

Sapinsley gives a short history of the establishment of seven agricultural colonies in Kansas. He focuses on the Beersheba Colony in Finney County, Kansas and the support provided by the Hebrew Union Agricultural Society. The Beersheba colony was established in 1883 with a range of about 175 to 200 families. Although they were welcomed by neighboring farmers who supported them and offered advice and assistance, the colony endured for about three years. Sapinsley points to a number of factors that led to the closing of the colony. Among them are unsuitable land and weather conditions, lack of farming skills, insufficient funds, disagreements over management, and settler dissatisfaction.

130. Scheler, Michael B. (1930 February). The "Back-to-Land" Movement of the Jews. *Reflex*, 6(1): 32–38.

Scheler asserts that Jews have a history of involvement in agriculture since biblical times but were barred from agriculture in the diaspora. He points to evidence of that involvement from back to the land movements during the 19th and 20th Centuries in Russia, Poland, England, South America, Asia and the United States where government and private organizations are providing financial and technical support for the settlements. He concludes that "back to the land" is a mass movement among Jews that will continue.

131. Schloff, Linda Mack (1996). "We Dug More Rocks": Jewish Women and Work in the Upper Midwest. *Minnesota History*, 55(4): 166–174.

Excerpts of oral histories from Jewish women in rural communities in the upper Midwest in the early 20th century from Schloff's book, *"And Prairie Dogs Weren't Kosher": Jewish Women in the Upper Midwest Since 1855* (Minnesota Historical Society, 1996). The women express their views on issues that affect their daily lives including farming, work, and entrepreneurial activity.

132. Schnapik, Ceil Pearl (1983). Jewish Farmers of the Benton Harbor Area. *Michigan Jewish History*, 23(2): 3–11.

The story of Benjamin Rosenberg and his family, the remaining members of a three generation farm family in Benton Harbor, Michigan. The Jewish agricultural experience in the Benton Harbor area (northern Berrien County) began in 1904. By 1950, about 20 families out of a population of about 250 families were engaged in agriculture in Sodus, Berrien Springs, Dowagiac and Eau Claire areas. It dwindled to one family, the Rosenbergs, still engaged in full time farming by 1981. Schnapik presents a history and genealogy of the three Rosenberg generations and brief descriptions of other former Jewish farm families in the area.

133. Schwartz, Lois Fields (1965). Early Jewish Agricultural Colonies in North Dakota. *North Dakota History*, 32(4): 217–232.

A large immigration of German Jews to the United States took place between 1815 and 1890. An extensive immigration of Russian and East European Jews occurred between 1881 and 1920. The German Jews settled in cities and assimilated readily into American society. When the Russian Jews arrived, some settled in cities and got involved in garment industries. The Jewish Agricultural Society offered economic aid to settle Russian Jews in agricultural colonies in North Dakota and other states. Schwartz describes living and working conditions on three colonies established in 1882. Painted Woods colony settled initially by 11 families northwest of Bismarck on the north shore of the Missouri River; Garske, located 15 miles north of Devils Lake, and the Iola colony 20 miles northwest of Devils Lake. By 1884 Painted Woods had 1400 cultivated acres, 53 horses, 56 oxen, 61 cows, 86 calves and some houses and barns. Despite attempts to make the colonies succeed, settlers faced numerous obstacles including lack of farm experience, bad weather, poor soil, economic hardships, disagreements about governance, and discrimination by neighbors. As a result, the colonies lasted only a few short years and were abandoned by 1889. Most of the settlers moved to urban areas and entered other professions. Contains excerpts of letters from settlers expressing their frustrations with the economic and social conditions in the colonies.

134. Shankman, Arnold (1978). Happyville, the Forgotten Colony. *American Jewish Archives*, 30(1): 3–19.

In 1905, Russian immigrant Jews attempted to establish an agricultural settlement called Happyville in Aiken County, South Carolina. The colony failed after two years for several reasons. The inhabitants had no former farming experience, finances and debts were a continual burden, and severe weather destroyed the crops. Shankman discusses the formation of the colony, the idealism of its inhabitants, and the challenges they faced.

135. Shaw, Albert (1908). Jewish Farmers. *American Review of Reviews*, 37(5): 617–618.
Brief description of the New Jersey agricultural settlements at Alliance, Norma, Brotmanville, Carmel, Rosenhayn, and Woodbine. Gives statistics on size, number of residents, kinds of enterprises, agricultural produce, and manufactured products.

136. Shilo, Margalit (1995). Sicily Island and Rishon Le-Zion: A Comparative Analysis. *American Jewish Archives*, 47(2): 182–212.
The mass migration of Jews from Eastern Europe from 1881–1914 to the United States and Palestine offered immigrants the opportunity to choose agriculture as a profession. Shilo compares the establishment of an agricultural colony of Russian Jews in Sicily Island, Louisiana (1881) and an agricultural moshav in Rishon le-Zion in Palestine (1882). Sicily Island received financial and moral support from the Jewish community in Louisiana but failed in less than a year due to economic, social, and climactic problems. The colony at Rishon le-Zion persevered and survived.

137. Shpall, Leo (1965). Korolenko's Letters on the Woodbine Colony. *Proceedings of the New Jersey Historical Society*, 83(2): 93–106.
Vladimir Korolenko, the well-known Russian writer and editor, visited the Jewish agricultural colony of Woodbine, New Jersey while on a trip to the World's Exposition in San Francisco in 1893. He was impressed with the way in which the settlers had adapted to life in a cooperative agricultural community and adjusted well to American life. He wrote three letters to his wife about his visit and impressions of Woodbine intended for publication in the Russian press. The letters never appeared due to censorship but were published in 1933 in a collection of Korolenko's letters. The three letters, translated to English, are published here in full.

138. Shpall, Leo (1950). Jewish Agricultural Colonies in the United States. *Agricultural History*, 24: 120–146.
A detailed history of the movement to settle Jewish immigrants on agricultural settlements in America. Shpall begins by citing settlement proposals as

early as 1783 and takes readers through a chronology of the plans and proposals through 1950. He analyzes the organizations that assisted with settlement projects, the major players in the settlement movement, and activities during the major immigrations from Poland, Russia, and Germany. Shpall discusses the establishment of colonies in Louisiana, Kansas, South Dakota, Oregon, Colorado, Kansas, and New Jersey. Some of the colonies failed after a short time, others were still in existence in 1950. He analyzes the economic, social, cultural, and religious factors that had an impact on a colony's success or failure.

139. Shpall, Leo (1937 July). A Jewish Agricultural Colony in Louisiana. *Louisiana Historical Quarterly*, 20(3): 821–831.

A history of the founding of the first Russian Jewish agricultural colony in the United States. In 1881, there was a large influx of Jewish immigrants from Russia seeking a better life in America in an agricultural setting. Shpall describes the efforts of various organizations to establish a colony for the immigrants in Sicily Island, Catahoula Parish, Louisiana in 1882. The Hebrew Foreign Mission Society, the Alliance Israelite Universelle and the Union of American Hebrew Congregations assisted in the effort and tried to recruit New Orleans Jews to assist in the settlement program. Settlers were enthusiastic about the Sicily Island project and even wrote a constitution with 17 articles outlining their governance and operating procedures. Nevertheless, a number of serious problems kept the colony from succeeding. Poor soil conditions, lack of farming knowledge, feelings of isolation, malaria, and floods led to the abandonment of the colony within a few years. Includes the "Constitution for the First Agricultural Colony of Russian Israelites in America."

140. Siegel, Thyme S. (1995 [5756]). The Lost Tribe. *Bridges: A Journal for Jewish Feminists and our Friends*, 5(2): 49–53.

Siegel reminisces about her childhood on a chicken farm in Lakewood, New Jersey near Farmingdale in a rural Jewish community of mainly immigrant Jews from New York who moved to Central New Jersey during the 1930s through the 1950s. She relates her personal thoughts and views about her Yiddish speaking immigrant parents, the farm community, Jewish land ownership, her secular and religious education and her views about God and religion. The community formed cooperatives and unions to serve the community and help them survive as farmers. The suspicions and pressures of the anti-communist hysteria of the 1950s pitted members of the Farmingdale community against each other resulting in devastating divisions and turmoil. Rifts developed among socialists and communists and so called "right" and "left" supporters. Siegel explores these rifts and offers her observations regarding its impact on the Jewish farmers in the region.

141. Solomons, Adolphus (1888 September). Alliance: The First Successful Jewish Colony in America. *Menorah*, 5(3): 179–187.

Solomons presents a short history of the Alliance Colony established in 1881 in Vineland, New Jersey. He explains how the Hebrew Emigrant Aid Society which was founded in response to the large influx of Russian Jews and the pressing need to encourage them to enter industrial and agricultural professions. Eight years after its initial settlement, Solomons visited the colony and found a mixed agricultural/industrial enterprise with successful crop yields and a profit making clothing factory. The 135 families at Vineland have homes with modern appliances, a synagogue, liberal education for the children, and a post office soon to be built. He concludes that Vineland is doing well and prospects for the future are encouraging.

142. Soltes, Mordecai (1924). Prevailing Attitudes on Outstanding Civic Problems. A: Urban Congestion — Urge the Jews to Return to the Soil. *American Jewish Year Book*, 26: 265–269.

Part of an article: "The Yiddish Press — An Americanizing Agency" (pp. 165–372) presenting excerpts from the Yiddish press on various topics. This segment contains articles extolling the values of agriculture and urging American Jews to return to the soil. Gives three examples of items from Yiddish newspapers supporting this view. (1) A description of a convention of Jewish farmers in New York from *Morgen Journal* wishing the farmers success and prosperity; (2) An editorial in *Tageblatt* about the importance of the education of children of Jewish farmers in both secular and religious subjects; (3) Response to a letter from New York to the editor of *Der Tog* written by a father whose son is enrolling in an agricultural college and wants to become a farmer. The editors offer good reasons why the father should support his son and encourage him to study at Cornell University.

143. Stern, Julius (1843 April). On the Establishment of a Jewish Colony in the United States. *Occident*, 1(1): 28–32.

Stern makes a case for the establishment of a Jewish colony in the Western Territories to encourage German Jews to emigrate to the United States. He envisions a mixed agricultural-industrial colony that will enable them to learn farming and business techniques and become respectable and independent members of American society.

144. Stern, Norton B. (1991). The Early Interest Which Preceded the Establishment of Jewish Farm Colonies. *Western States Jewish History*, 23(3): 223–230.

Although the first Jewish farm colonies in the United States were established in 1882, interest in this idea began in the late 1870s. Lazard Kahn proposed the idea in a letter to the *Cincinnati Jewish Press* in 1879. An article in the *New York Jewish Weekly* reported on emigrants from Hungary that were desirous of settling in a colony on government land. In the same year, the editor of *Pittsburgh Gazette*, a secular newspaper, asserted that settlement would have a "beneficial effect on the public's opinion of Jews." At the Sixth Annual Meeting of the Union of American Hebrew Congregations (HUAC), Moritz Ellinger of New York and Moritz Loth of Cincinnati recommended that a substantial sum of money be allocated to establish a successful farm colony. The proposal was not acted upon by HUAC. Two years later in 1882, several colonies were established with support from other sources.

145. Stern, Norton B. (1982). Colorado Homesteader and Los Angeles Investor: A Picture Story. *Western States Jewish Historical Society Quarterly*, 14(4): 302–307.

Henry Jacob Clar was the son of Russian Jewish immigrants who came to America in 1885. After living in New York for several years, the Clars moved to Denver where Henry grew up. He opened a shoe repair shop with his brother Sam in 1905. In 1906 Clar obtained homestead land in Morgan County, Colorado and began farming. In 1913 Clar and his wife received a land patent to farm 320 acres in Morgan County. Clar's uncle and brother in law joined in the venture by obtaining adjoining homesteads. Clar continued to farm until 1922 when he leased the land to tenant farmers and left to settle in Los Angeles. Oil was discovered on Clar's land in 1953 and Union Oil company began drilling. Clar invested in property in Los Angeles and later bought an additional 5,000 acres of farmland in Colorado. This article is based upon an interview with Clar's sister Elizabeth and includes various photographs from the family's collection.

146. Stern, Norton B. (1982). The Jewish Dairyman of San Francisco. *Western States Jewish Historical Quarterly*, 14(2): 167–174.

A profile of Isaac Stone (1825–96), who operated a dairy with his wife Hannah and seven children (five sons, two daughters) two miles south of San Francisco, starting in the 1860's. The Stone dairy received San Francisco Milk Permit No. 1. Stone had 175 cows and delivered milk to customers in the San Francisco area. The Stone sons worked in the dairy, drove delivery wagons, and helped manage the business. However, only one son, Abraham Lincoln Stone remained in the family business and operated the dairy after his father's death. He also acquired a ranch in the San Joaquin valley. The dairy changed names several times and undertook several mergers. The last merger was with the Borden Company in the mid 1920s, which was then a San Francisco company. Abraham Lincoln Stone was recognized as one of the leading dairy experts on the Pacific Coast. He died in 1931.

147. Stern, Norton B. (1978). The Orangevale and Porterville, California Jewish Farm Colonies. *Western States Jewish Historical Quarterly*, 10(2): 159–167.

The International Society for the Colonization of Russian Jews was organized in 1891 by David Lubin to settle Russian immigrant Jews on the land. Lubin and his half brother, Harris Weinstock, settled 10 families on land they owned at Orangevale, several miles northeast of Sacramento. Weinstock operated a water company and about eight farmers were employed by the company. The land was unproductive and the colony was abandoned three years later. Philip Nettre Lilienthal, a banker and land developer, established the Pioneer Land Company and acquired 6,400 acres near Porterville in 1891. He arranged for 15 families to settle on the land and raise alfalfa, fruit, vegetables, and poultry. The experiment was abandoned within three years. The failures of the Orangevale and Porterville colonies were due to poor land, inadequate capital, lack of farming skills, and low motivation to pursue a life in agriculture. Stern points out that although many Jewish agricultural settlements in Western states did not last, there were many Russian Jewish farmers in California who operated independently and were economically successful.

148. Stern, Norton B., and William M. Kramer (1981). An American Zion in Nevada: The Rise and Fall of an Agricultural Colony. *Western States Jewish Historical Quarterly*, 13(2): 130–134.

In June 1897, Morris Cohn, a real estate promoter, convinced 25 newly arrived Russian Jewish immigrant families living in San Francisco to establish an agricultural settlement in Smith Valley, Nevada on a large ranch consisting of 5500 acres. When it became apparent that the settlers could not purchase the tracts for the $26,000 price that Cohn wanted, it was purchased by a Nevada pioneer, Samuel Wymore, who sold the land to the settlers with houses, cattle and implements for $12,000 to be repaid in ten equal installments. The settlement was opposed by Rabbi Jacob Voorsanger of Congregation Emanu-El in San Francisco who claimed that the settlers were being used for a business deal that would not succeed. As predicted, the colony failed one year later and the land was sold. The reason for failure included opposition to the cooperative nature of the colony, lack of irrigation water, undercapitalization, a bank foreclosure, and the attraction of small businesses in nearby communities which had greater appeal to the immigrants.

149. Stine, Andy (2003 Autumn). Suffering from Want: The Jewish Colony at Cotopaxi. *Colorado Heritage*, pp. 38–47.

Story of the Jewish agricultural colony at Cotopaxi settled by about 60 Russian immigrant families seeking an alternative to life in the city. The colony was established in 1882 with the assistance of the Hebrew Immigrant Aid Society and Emanuel Saltiel, a wealthy mining and milling entrepreneur who offered land for the settlement and agreed to assist with housing, implements, and supplies for the settlers. The article describes the settlers' backgrounds and conditions at Cotopaxi. Stine alleges that the settlers were poorly treated by Saltiel who failed to fulfill his obligations. The settlers had to find work in Saltiel's mine or with the Denver and Rio Grande Railway to support themselves and pay their debts. Although the land was not suitable for farming, the settlers persevered and managed to get small yields of potatoes and corn which was completely destroyed by a late spring blizzard. By June, 1884, the colony was dissolved and settlers left. Some operated successful farms elsewhere, others became ranchers or business people and some moved to Denver and entered business or professions.

150. Swichkow, Louis J. (1964). The Jewish Agricultural Colony of Arpin, Wisconsin. *American Jewish Historical Quarterly*, 54(1): 82–91.

A small agricultural colony of 720 acres was established at Arpin, Wisconsin in 1904. Adolph W. Rich, a Milwaukee merchant, manufacturer, and philanthropist, organized the colony with the assistance of the Milwaukee Jewish Industrial Aid Society. The colony began with five families and reached fifteen families (about 75 people) at its peak in 1906. The settlers faced inclement weather, hard work and low income but they survived by working cooperatively with each other and with Christian farm neighbors. They preserved their religious observance and traditions and their dream of building a synagogue and providing Jewish education for their children was finally realized in 1915. However, the lack of secondary educational facilities, minimal social contacts with other Jews, and the fear of intermarriage led to the dissolution of the colony in 1920. As the children grew up, they left Arpin to settle elsewhere and were followed by their aging parents who were no longer able to farm.

151. Teasdle, Holly (2002 Fall). Jewish Farming in Michigan. *Michigan Jewish History*, 42: 2–14.

Presents a history of Jewish agricultural settlement in Michigan from m the 1880's to the 1930's. There were settlements in Bad Axe, Alicia, Benton Harbor, Bangor and other areas of the state. Many of the settlements were short lived and most were gone by 1938 due to scarcity of money and resources, lack of agricultural experience, political differences and other factors. Some farm families remained in business and operated successful fruit, vegetable, and livestock enterprises in Ann Arbor, Manistee, Upper Peninsula and Rochester. By 1950 the number of Jewish farmers in the State declined along with the decline in the general farming population. Teasdle profiles prominent Jewish farmers and their role in Michigan agriculture. (Also available online at: http://www.michjewishhistory.org/pdfs/vol42.pdf.)

152. Trauner, Ruby (1995 [5756]). Working Class Words on Tenements and Gardens. *Bridges: A Journal for Jewish Feminists and Our Friends*, 5(2): 56–67.

Three Jewish women who grew up in apartment buildings in New York City and now live in Seattle present their views on Jewish attitudes regarding class and land. They talk about how the concepts of land entered their lives through urban gardening, summers in upstate New York, and relatives who farmed. They give their observations about the impact of land and land ownership on Jews in rural and urban communities.

153. Vanderpoel, Sally (1997). Jewish Agricultural Society Experiment in Eastern Wyoming, 1906–1918. *Annals of Wyoming*, 69(3): 2–10.

Vanderpoel, whose family came to the Village of Huntley, Wyoming in 1922 to settle on homestead land, presents a historical narrative about pioneering Jewish families who acquired homesteads in Eastern Wyoming and tried to make a life on the farm in the early 20th Century. In 1906, the Jewish Agricultural Society established a farming settlement in the region with fifty six Jewish families from New York and other states. The area was undergoing irrigation and land reclamation projects but that did not alleviate the settlers' struggles. They suffered economic hardship, scarce resources, unsuitable soil conditions, and lack of agricultural skills. By 1918 most of the original settlers had left; a few remained on their land. The Harry Fieldman family was one of the first settlers. Other early settlers were the DeBrots, Yocums and Mortons. Fieldman's son, Frank, took over the farm in 1922; he left in 1942 and settled in Cheyenne. Drawing upon reminiscences handwritten by Harry Fieldman's son, Frank, and interviews with local individuals, Vanderpol describes life on the Wyoming prairie and the hardships involved in making the land productive.

154. Waterman, Sigismund (1855 May). A Call to Establish a Hebrew Agricultural Society to Encourage Agriculture Amongst the Israelites of America. *Asmonean*, 12(2): 22.

A proposal to establish the "American Hebrew Agricultural Association" to create an interest in agricultural pursuits for immigrants from Eastern Europe. The Association will be chartered under the laws of the State of New York and governed according to a constitution. The proposal recommends the purchase of land in New York State to produce agricultural commodities for the New York metropolis and establishment of a school to train those interested in farming.

The project failed to gain widespread support and was abandoned. (Also published with minor variation and comments as: *Jews in Farming* in: Schappes, Morris U, Editor. A Documentary History of the Jews in the United States, 1654–1875. New York, Shocken Books, 1950, pp. 346–349.)

155. Weinryb, Bernard D. (1954). Noah's Ararat Jewish State in Its Historical Setting. *Publications of the American Jewish Historical Society*, 43(3): 170–191.

An analysis of two plans drawn up in 1819 regarding the establishment of settlements in the United States for Jewish immigrants fleeing persecution and economic deprivation in Russia. Weinryb gives details of Mordecai Noah's efforts to establish Ararat, at Grand Island, New York and William Davis Robinson's *Memoir addressed to persons of the Jewish religion in Europe on the subject of emigration to, and settlement in, one of the most eligible parts of the United States of North America.* Weinryb examines the motivations and philosophies behind both plans and the attitudes of the time favoring immigration. Robinson failed to gain widespread support for his plan from European or American Jewry. Noah envisioned a community with agriculture and light industry which would become self sufficient and allow the immigrants to integrate into American life. Although Noah had supporters in the United States and Europe, the plan was widely criticized by Jewish leaders. Noah abandoned the plan and moved on to other ventures. He became a Zionist and continued to advocate for Jewish deliverance and focused his efforts on a return to Palestine. (Also published in: Karp, Abraham J., editor. *The Jewish Experience in America. Volume II—In the Early Republic.* Waltham, MA., American Jewish Historical Society, 1969, pp. 136–157. See also: Robinson, William Davis [1918]. *Memoir addressed to persons of the Jewish religion in Europe on the subject of emigration to, and settlement in, one of the most eligible parts of the United States of North America.*)

156. Werner, Alfred (1948). American Jews on the Land. *South Atlantic Quarterly*, 47(3): 342–351.

Werner traveled to several parts of the country observing Jewish farm communities and interviewing former settlers (whom he called "patriarchs") about their lives and work. He conducted the interviews with patriarchs from Jewish farm communities in Benton Harbor, Michigan, Vineland, New Jersey, Colchester, Connecticut, and Sholem Colony in Wawarsing, New York. The patriarchs provided stories about their lives on the farms and their joys and frustrations. Werner offers his observations about the Jewish "back to the land" movement and the support of the Jewish Agricultural Society, Baron de Hirsch Agricultural School and the newspaper *The Jewish Farmer* that provided economic support, education and information to help the farmers succeed.

Newspapers

Special Newspapers Serving Jewish Farmers

157. *The Jewish Farmer* (1900–1959). New York: Jewish Agricultural Society. Published Monthly.

The Jewish Farmer, which described its mission as "a journal devoted to the interests of agriculture among the Jews in the United States and Canada," was published monthly in English and Yiddish and served as the official voice for Jewish farmers for five decades. Each issue contained practical advice on fruit, vegetable, and crop production, livestock and poultry raising, sanitation, farm safety, farm management, news about agricultural developments in the United States and Israel, and full coverage of the Jewish Agricultural Society and its activities, conferences, committees, and reports. Articles documenting the progress and contributions of Jewish farmers to American agriculture appeared regularly. For a history of the newspaper and its impact on the Jewish farm scene, see Miller, Benjamin, "50 Years of The Jewish Farmer," in the May 1958 issue. The final issue was published in December, 1958 with a sad farewell to its readers. Changing times, diminishing circulation, rising costs and competition with other information sources led to a decision to cease publication. The final issue includes a memorial to Dr. Gabriel Davidson, the Executive Director of the Society, who died earlier in the year.

158. *Sunrise News* Alicia, Michigan: Sunrise Co-operative Farm Community (April 1934–May, 1935).

This newspaper served the Sunrise Cooperative Farm Community during its first year. It contained information and news of interest to the members, summary of weekly meetings, general articles on politics and national issues, and school news.

159. *Rural Roots: Jewish Farm History Newsletter* (Vol. 1, 1989–Vol. 7 1995). Princeton, NJ: Documentary III.

Documentary III is an organization working to preserve the history of Jewish agriculture in New Jersey. Their publication, *Rural Roots*, was a medium of communication for former members of New Jersey farm colonies and independent farmers and their families. Readers contributed anecdotes, stories and photographs. The Newsletter also featured public events such as films, exhibits, and guest columnists to keep the spirit and historical awareness of Jewish agriculture in America alive.

General Newspapers

160. Adler, Cyrus (1884 August 8). The Russian Jews at Alliance. *American Hebrew*, 19 (13):200–202.

The Alliance colony was established in May, 1882 on 1100 acres with seventy families comprising 750 people. Adler visited the Colony and offered his observations and thoughts about its future. At the time of his visit, it contained fifty families with about 250 people. In addition to agriculture, the colony had a cigar factory that produced about 7000 to 9000 cigars a day. Adler describes how the land was acquired, the street layout, housing, economic situation, settlers needs, the soil and land conditions and problems that needed to be addressed. Adler observed that the settlers are motivated and believe that it will be a successful venture.

161. Allen, H. B. (1940 November 1). The National Farm School. A Way Back to the Soil for the Jewish Boy from the City. *American Hebrew*, 127 (25):7, 14.

The National Farm School was founded in 1896 in Doylestown, Pennsylvania to provide city youth and particularly Jewish youth with an alternative to city life by offering training in practical agriculture that would enable them to enter farming as a livelihood. Allen outlines the school's objectives, admission requirements, programs of study, and practical experience.

162. Anonymous (1943 May 9). Jews Thrive on U.S. Farms by Thousands. *Washington Post.*

There are more than 100,000 hardworking Jewish farmers in the United States in all areas of agriculture; many have contributed to wartime food production programs. Some of the distinguished achievements of these Jewish farmers include the largest potato grower in Maine; the country's largest onion grower; a former immigrant runs an 8000 acre farm in California; the largest milk producer in New Jersey; and a poultry breeder in New York, who raised a prize bird that was the world's largest egg producer for five years. A Jewish farmer became president of the South Dakota seed growers association and another was offered the senatorial nomination of the Farmer-Labor Party (he declined).

163. Anonymous (1936 August 7). Colony Nearing Its Goal — Cohen. Quiz Sunrise Unit Leader. *Saginaw Daily News*, pp. 1–2.

Joseph J. Cohen, chief defense witness in the suit to stop receivership of the Sunrise Co-operative Farm Community, acknowledged that there were dissident members who felt that revenue received was not distributed fairly and that they were not adequately compensated for their skills. He testified that the Colony is functioning more smoothly now than at any time in its three year existence.

164. Anonymous (1936 June 1). U.S. Takes Up Option on Sunrise. May Be Cut Up in Small Units. *Saginaw Daily News* (Saginaw, MI), p. 1.

The final step in dissolution of the Sunrise Cooperative Farm Community was taken this week by the Rural Settlement Administration (RSA). RSA will purchase the 9000 acre farm for $68.000. It will be divided up into a number of small farms for public sale. The seventy families now living in the Sunrise Community are expected to remain on 1500 acres.

165. Anonymous (1933 October 1). Community Farm Plans for Winter. Activities in the Community of 200 Residents. *Saginaw Daily News* (Saginaw, MI), p. 2.

The Sunrise Co-operative Farm Community is continuing to build and plans to have a successful year. Members are building a community kitchen to use for communal dining services. A high school is being established to educate fifteen youngsters who have reached high school age. Work in the fields is drawing to a close as winter approaches. The members remain motivated by the cooperative spirit and are looking forward to a good future.

166. Anonymous (1935 February 15). Jews Are Seen Turning to Farm. Agricultural Society Report Cites the Importance of Rural Gains for Them. *New York Times*, p. 12.

Summary of the 1934 annual report of the Jewish Agricultural Society. Gabriel Davidson, General Manager of the Society states that the notion that Jews are unfit for agricultural pursuits is not valid. Loans were granted to 381 applicants and twenty per cent of the loans advanced in 1934 have already been paid and more will be repaid by the end of 1935.

167. Anonymous (1934 February 1). Colony Fails at Farming, but Gets Riches from Oil. *Wall Street Journal*, p. 12.

A farm colony of Jews with sixty families near Liberty, Texas midway between Conroe and Cleveland, failed some years ago and was abandoned. The owners of the tracts are now leasing their holdings for oil and gas development. The tracts run from 44 to 176 acres and have been leased for sums as high as $18,000. If oil is found, each owner will receive one eighth royalty on all production.

168. Anonymous (1932 May 6). Gabriel Davidson Feted on 25th Anniversary. *Mountain Hotelman* (Ellenville, NY), 3 (6):1–2.

Gabriel Davidson, General Manager of the Jewish Agricultural Society, was honored at a testimonial dinner celebrating 25 years of service. The event was hosted by directors and staff of the Jewish Agricultural Society and the Baron de Hirsch Fund. Guest speakers included Jacob G. Lipman, Dean and Director of the New Jersey Agricultural College (Rutgers University), and Mrs. Elmer Eckhouse, Executive Director of the National Council of Jewish Women. Congratulatory messages from governors, members of Congress and the Secretary of Agriculture were received. Davidson was commended for his understanding of the needs of Jewish farmers and the Society's role in supporting them.

169. Anonymous (1932 February 29). Topics of the Times: A Unique Philanthropy. *New York Times*, p. 16.

A tribute to the work of the Jewish Agricultural Society in honor of Baron Maurice de Hirsch on the 100th anniversary of his birth. (Hirsch died in 1897.) In the last five years, more than 300 families have been established on farms with the Society's support. Since its founding, the Society has given 10,288 farm loans totaling nearly $7 million.

170. Anonymous (1931 November 20). Federation News. *Mountain Hotelman* (Ellenville, NY), 2 (20):1,4.

Gabriel Davidson, General Manager of the Jewish Agricultural Society, spoke at the 2nd Annual Convention of the Federation of Hotelman's Associations. He urged the adoption of a resolution for greater cooperation among farmers and hotelmen by encouraging hotels to buy their produce from the farmers whenever possible. A former farmer who was attending the convention indicated that buying their produce provides only a small amount of business. The hotels should contract with the farmers to supply them with products directly from their farms.

171. Anonymous (1931 March 27). Jewish Farmers Weather Depression. *Mountain Hotelman* (Ellenville, NY), 2 (3):7.

Summary of the 31st annual report of the Jewish Agricultural Society describing the work and progress of the departments and branch offices. There are over 80,000 Jewish farmers on over 1 million acres throughout the United States. Gives statistics and details on the activities of the settlement, extension, and sanitation departments. A recent study of a cross section of Jewish farmers in thirteen states indicated that more than half have been on their farms over ten years and a third over fifteen years. Ninety percent of the farmers in the survey intend to remain on the land and sixty four percent hope their children will engage in an agricultural enterprise.

172. Anonymous (1931 July 3). Sidney Levine, New Extension Agent. *Mountain Hotelman* (Ellenville, NY), 2 (20):1.

Report on the appointment of Mr. Sidney Levine as extension agent in the Ellenville area. Levine was raised on a poultry farm in Woodbine New Jersey and studied agriculture at Rutgers University graduating with high honors. Levine brings to the position "the finest tradition of rural family relationship as well as a thorough understanding of agricultural problems."

173. Anonymous (1928 April 15). Jewish Farmers Now Till a Million Acres. Holdings of 80,000 Are Valued at $150,000,000, the Jewish Agricultural Society Reports — How They Are Aided and How They Are Succeeding. *New York Times*, p. 134.

Describes the work and accomplishments of the Jewish Agricultural Society on its 28th year of operation. Gabriel Davidson, the Society's General Manager, explains the long involvement of Jews in agriculture and the Society's mission to assist Jewish immigrants get established in farming. He discusses the society's goals, the financial assistance, the millions that have been loaned, and achievements of prosperous Jewish farmers around the country.

174. Anonymous (1927 July 1). Back to Mother Earth. The Weil Family Discover the "Promised Land" in Old Kentucky. *American Hebrew*, 121 (8):285–286.

Story of the agricultural accomplishments of the Weil family in the Kentucky Bluegrass region. Jonas and Sim Weil emigrated from Alsace-Lorraine in 1887 and became active in the livestock business. Their farm operation grew from 324 acres in 1897 to 3300 acres in 1917. The Weil farm raised cattle, sheep, lambs and high yields of bluegrass seed, wheat, rye, corn, hay and burley tobacco. After a thirty year partnership they separated and continued to prosper on individual farms of over 2000 acres. The Weils were famous throughout Central Kentucky as successful farmers and experts who were always available to offer advice and assistance to other farmers in the area.

175. Anonymous (1926 May 9). Want City Youths to Take Up Farming. *New York Times*, p. E19.

Report of the National Farm School Conference held at the Biltmore Hotel in New York. Politicians, agriculture officials, and supporters of the school attended. The purpose of the six day conference was to attract young men and women from the cities to learn agriculture and prepare for careers in farming. Briefly describes the school and its grounds and the kinds of youngsters they hope to attract.

176. Anonymous (1923 June 8). Farm School Exercises. *Jewish Tribune*, p. 16.

A large gathering attended the annual spring exercises of the National Farm School in celebration of the 25th anniversary. It was announced that $600,000 had been raised toward a million dollar endowment fund. The Nathan Straus Vineyard was dedicated in honor of the 75th birthday of Rabbi Joseph Krauskopf, the school's founder. Edward C. Stokes, the former Governor of New Jersey (1905–1908), attended the exercises with a delegation of 120 residents of Newark who raised $35,000 for the school. Stokes indicated that agriculture is a part of the Jewish legacy. He said that Jews were the first scientific agriculturists and among their innovations was the introduction of tobacco into the American colonies. In addition, the experience of the agricultural colony at Woodbine attests to the ability of Jews to become tillers of the soil.

177. Anonymous (1914 February 13). From Sweatshop to Ranch: Back to Soil By Colonies. Jews of This City Interested in Co-Operative Plan. *Los Angeles Times*, p. 117.

Benjamin Brown, President and General Manager of Clarion, the Jewish agricultural colony near Gunnison, Utah, visited Los Angeles and San Francisco seeking support for establishment of a colony in California. Some of the bonds sold to finance the Utah colony were sold in Los Angeles and San Francisco and Brown is seeking colonists and economic support for a California colony similar to the Utah model. He describes Clarion and points out that it was established with the support of the Jewish Agricultural and Colonial Association which plans to establish farm colonies with 100 or more families in several states. Wealthy Jews in various parts of the country buy the bonds at 4 percent interest. These bonds help finance operating and other expenses on the colony. Brown pointed out that there are 47,000 Jews on farms in the United States and two Jewish agricultural schools educating sixty to seventy five graduates yearly. To join the colony members must possess $300, be a U.S. citizen, and be willing to live and work in the settlement.

178. Anonymous (1911 October 9). Jews Plan Big Farming Colony. Chicago Leaders Form "Philanthropic Syndicate"; Buy Wyoming Tract. *Chicago Daily Tribune*, p. 9.

Arrangements have been completed to resettle 450 Jewish families on a 72,000 acre agricultural settlement in Wyoming about seventy miles north of Cheyenne known as the Wheatland district. The settlers are poor immigrants recruited from ghettos and sweatshops who are willing to pursue a life on the land. The colony is funded by a grant of $450,000 from a "philanthropic syndicate" headed by Julius Rosenwald. If the colony succeeds, the settlers will have 15 years to pay for their farms. The Immigration Commissioner of Wyoming extended a warm invitation to the settlers and offered to provide experts to advise on farming techniques.

179. Anonymous (1911 June 29). Jews Plan Farming Colony. Fifty Chicago Families Will Go to Minnesota. *Chicago Daily Tribune*, p. 14.

Fifty Orthodox families from Chicago will be leaving the City to form a communal agricultural colony in the timberlands of Northern Minnesota. Each family will buy 40 acres with economic support from a fund set up by Minnesota Governor A.O. Eberhart.

180. Anonymous (1909 October 8). Farm School's Annual Meeting. *Jewish Exponent*, 49(26):9.

Report of the activities at the Succoth Pilgrimage to the National Farm School and the 12th Annual Meeting of the National Farm School Association. The event included a talk by Rev. Dr. Joseph Krauskopf, Founder and President of the Farm School, Samuel S. Fleischer, Chairman of the Baron De Hirsch Agricultural School at Woodbine, and John H. Washburn, Director of the Farm School, regarding the education and training programs and its impact and benefits to Jewish farmers around the country. Treasurer's report and student awards were also presented.

181. Anonymous (1909 October 1). What Becomes of Agricultural Students. *Jewish Exponent*, 49(25):10–11.

Selected letters from former students of the Baron de Hirsch Agricultural School at Woodbine (NJ) describing their work and the influence the school had on their lives.

182. Anonymous (1905 November 20). Wants Land for Jews. New Yorker Has Plan to Raise $500,000,000. *Chicago Daily Tribune*, p. 3.

Dr. M.J. Burstein of New York City appeals to all Jews in America and Europe to give 10 percent of their possessions to buy land for Russian Jews who are suffering under a cruel regime. He proposes that every Jew give up 10 percent of his capital or investments at once and 10 percent of yearly earnings for the next ten years. Burstein wants to raise $500,000,000 to purchase 5 million acres of land in Texas or the territories which he believes can be paid for in fifty years. The money will be used to buy the land, transport the Russian Jews to the United States, provide farm implements and machinery, and build homes, schoolhouses and public buildings. After only nine days, the relief fund has already raised $500,000 from Jewish philanthropists and other sources.

183. Anonymous (1902 January 27). New Jersey Jewish Colonies: Their Part in Redeeming the Wild Lands of the State. *Atlanta Constitution*, p. 7.

Discussion of the four New Jersey Jewish agricultural colonies of Alliance, Rosenhayn, Carmel, and Woodbine. Briefly describes how each colony was established, the number of settlers, and role of the Baron de Hirsch Fund and other aid organizations.

184. Anonymous (1897 June 13). How They Teach Practical Farmer. *Atlanta Constitution*, p. 3.

Announces the opening of the National Farm School on June 14, 1891 near Doylestown, PA, twenty five miles from Philadelphia. Rabbi Joseph Krauskopf, founder of the school, raised the necessary funds and planned the school, its grounds, and facilities. The purpose of the school is to train young Jewish men and woman to be farmers, gardeners, orchardists, and livestock raisers. With the New Jersey agricultural colonies taking hold, Krauskopf hopes the school will prepare leaders for important positions in agriculture. Describes the admission requirements, courses of study, and how the school will function.

185. Anonymous (1897 June 13). Jewish School for Practical Farming. *Los Angeles Times*, p. 18.

Report on the opening of the National Farm School founded by Rabbi Joseph Krauskopf, a prominent Jewish leader, on 122 acres near Doylestown, Pennsylvania. The school was designed as a place to educate Jewish men and women for a career in agriculture. Although predominantly Jewish the school accepts all students who meet the admission standards. The 122 acre site contains the school, a farm with a dairy, greenhouses, hothouses, and other buildings and living accommodations. Krauskopf describes the curriculum and course of study and points out that agriculture is inherent in Jewish life from biblical times to the present. As examples he points to thriving Jewish farmers in Palestine, Russia, Argentina, and New Jersey.

186. Anonymous (1893 March 16). A Contented Colony. The Jews at Chesterfield Are Not Starving as Reported. *New York Times*, p. 2.

In response to reports in some New York newspapers, stating that the agricultural colony at Chesterfield, Connecticut is failing and settlers are destitute, Arthur Reichow, a representative of the Baron de Hirsch Fund, visited the colony. He disputed the reports by pointing out that the colony is doing well and only 6 of the 32 families are in need of assistance. Although the colonists were emerging from their first winter in the colony, the situation was not serious. A new creamery was built during the summer and the colony has 180 cows producing milk. Reichow will make arrangements to provide financial aid to the six families in the colony who need it.

187. Anonymous (1891 July 19). Baron De Hirsch's Fund. A New Russian Jew Colony in New Jersey to Be Located. *Chicago Daily Tribune*, p. 26.

Announces the plan by the Agriculture Committee of the Baron De Hirsch Fund to begin establishing agricultural colonies in the United States. A tract of land of about 5,000 acres has been acquired near Vineland, New Jersey for a mixed agricultural/industrial colony. The colony will begin with fifty families engaged in fruit and truck farming and a skirt factory. The plan will allot fifteen acres to each settler along with $700 cash and title to the land with fifteen years to pay for their plots. Professor Hirsch Leib Sabsovich, formerly of the College of Agriculture at Colorado State University in Fort Collins, CO, will serve as advisor. The colonies of Alliance, Carmel, and Rosenhayn were established earlier with the support of the Hebrew Emigrant Aid Society.

188. Anonymous (1891 July 11). Exiled Jews in New Jersey. *New York Times*, p. 4.

Examines the plan by Baron de Hirsch to establish mixed agricultural/ industrial colonies in Southern New Jersey for Russian immigrants. The article raises objections to the plan by pointing out that Jews have become assimilated into American society and have retained their own form of religion and worship. The agricultural colonies would create separate and segregated communities that raise moral and political questions.

189. Anonymous (1890 December 1). Aiding Jewish Exiles. A Movement Looking to Their Colonization in Free America. *Washington Post*, p. 2.

At a recent meeting of the Baltimore Branch of the Jewish Alliance, the group offered to assist Russian immigrants who come to the area to settle on farms. The Chair of the meeting, Henry W. Schneeberger, stated that this will enable those who wish to farm to get out of the crowded cities and live on the land. Pointing to the success of the New Jersey colonies, Schneeberger believes that the Jew can be a farmer when he is permitted to farm without restrictions. He asked for the cooperation of all who support the effort.

190. Anonymous (1883 July 20). U.A.H.C. The Closing Day. Big Work Done. *American Israelite*, 30 (3):4.

A report of the discussions, reports and votes on issues at the closing day of a meeting of the Council of the Union of American Hebrew Congregations. A proposal by the Committee on Agricultural pursuits favored the establishment of agricultural colonies in the United States. The Committee will have the power to raise funds for agricultural aid which will then be turned over to UAHC for agricultural purposes.

191. Anonymous (1883 March 17). Russian-Jew Farmers: Their Industry, Peculiarities and Religious Views. *Chicago Daily Tribune*, p. 6.

Copy of a letter originally sent to the editor of the St. Paul *Pioneer Press* regarding the Cremieux and Bethlehem-Yehudah Jewish agricultural colonies in North Dakota. The colonies, founded in 1882 by Michael Heilprin of New York City, have about seventy settlers and more are expected to arrive. The settlers are immigrants from Russia who have been in America two to three years. The writer visited the colony and describes the living arrangements, the school, and the economic arrangements. He commends the settlers for their perseverance, their willingness to work hard and their ability to save money and avoid incurring debts. He notes that when not engaged in their farming tasks, they are an educated people who are fond of reading, music, dancing, and politics.

192. Anonymous (1883 February 7). The Cotopaxi Colony. Sixty-Three Russian Refugees Suffering from Want in the Mountains. *Denver Tribune*, p. 6.

Reports on the arrival of the thirteen families who will settle at the Cotopaxi Colony raises questions about the decision of the Hebrew Emigrant Aid Society (HEAS) to establish a colony at that location. The writer states that HEAS offered each settler a house and furniture, cooking utensils, farm implements, seeds, and 100 acres of land as stated in a report by Julius Schwarz, President of HEAS, but this contradicts the conditions that actually exist at Cotopaxi. The land is unsuited for farming, the settlers were not supplied with enough implements, and have no agricultural experience. In addition, they arrived with scant clothing and provisions, the houses are too small and poorly constructed, and settlers have not been given enough funds for basic needs. Recommends that HEAS abandon the project and move the colony to a location where the settlers will be eager to farm and be rewarded with crop yields that bring income to the settlement. (See response to the allegations in: Denial from Cotopaxi. *Denver Tribune,* February 13, 1883, p. 8. See also: Schwarz, Julius. *Report of Mr. Julius Schwarz on the Colony of Russian Refugees at Cotopaxi, Colorado Established by the Hebrew Emigrant Aid Society of the United States, 1882.* New York, Hebrew Emigrant Aid Society, 1882.)

193. Anonymous (13 February 1883). Denial from Cotopaxi. An Unknown Correspondent Accuses the Denver Relief Committee of Misrepresentation. *Denver Tribune*, p. 8.

This response to the article "The Cotopaxi Colony" (February 7, 1883) regarding the conditions at Cotopaxi claims the article is at great variance with the facts and a great injustice to the people of Cotopaxi and to Julius Schwarz, the manager of the Colony. The anonymous author attempts to set the record straight regarding the funds, housing, productivity of the land and soil, and behavior of the settlers. The author states that the Committee that investigated the conditions at Cotopaxi got their information from the settlers who were inclined to make the case look bad. (See: Anonymous. The Cotopaxi Colony. Sixty-three Russian refugees suffering from want in the mountains. *The Denver Tribune*, February 6, 1883, p. 8.)

194. Anonymous (1879 July 11). Colony Farming for Jews: A Project Favored by the Hebrew Convention. *New York Times*, p. 2.

Report of a debate at a meeting of the Union of American Hebrew Congregations recommending the purchase of land to accommodate fifty immigrant families to enter the occupation of agriculture. Funds will be raised through private subscription or donations in co-operation with other organizations that support colonization. The families will have the privilege of paying for their land in seven years and can then request additional funds for more land under the same terms. The report was adopted with one dissenting vote.

195. Belluck, Pam (2002 November 1). For Orthodox Jews, an Experiment in Farming and Faith. *New York Times*, p. A24.

A group of Chabad Lubavitch Jews established a 70 acre kosher, organic, communal farm at Sunderland, Massachusetts. The farm, known as "Eretz HaChaim" (The Living Land), is designed to be a self sufficient agricultural community with a synagogue, Torah study center, schools, and a Mikvah (ritual bath). Seven families have moved to Eretz HaChaim; about twenty five families are expected. Non-Chabad religious and secular Jews are invited to join if they are willing to live as religious Jews. The farm plans to grow produce, make maple syrup, and raise sheep, goats, chickens and cows. Income will be generated from the sale of eggs, goat cheese, and organic chickens. The farm will operate under Jewish law and follow rules of the Torah including no work on the Sabbath (non-Jews will be hired to milk cows on Saturday), leaving corners of the fields for the poor, letting land fallow in the seventh year, letting orchards remain for three years before picking fruit, not planting certain fruits and vegetables next to each other, and donating food to charities. The founders hope that Eretz HaChaim will serve as a model for establishment of other Jewish agricultural communities in America.

196. Berney, Louis (1997 May 9). Over at the Star-K Ranch: The "Baltimore Chop" Has New Meaning in a Colorado Cattle Town Where

They're Cutting Certified Kosher Steers. *Baltimore Jewish Times*, 235 (2):28.

When economically depressed Mormon and Spanish ranchers in Alamosa, Colorado were looking for a niche product that would enable them to remain in farming, it was suggested that they consider the kosher market. A Denver Rabbi put them in touch with Rabbi Meyer Kurcfeld, who runs a kosher certification outfit called Star K. Kurcfeld visited the ranchers to educate them about the procedures necessary for proper kosher slaughter. Kurcfeld immediately won them over and the venture became a success. Today, about 60 ranchers operating as a co-op slaughter nearly 100 steers and 100 lambs a week primarily for the Los Angeles market. After two years of kosher production the ranchers are economically sound and the demand has increased so fast that they have trouble filling all orders. (For a documentary about this story, see the video *Kosher Valley*, Directed by Charles Davis. Boulder, CO, Throughline Productions, 1998.)

197. Davidson, Gabriel (1940 November 1). The Progress of the Modern Jewish Farmer. Forty Years of Service in the Jewish Farm Movement. *American Hebrew*, 147 (25):6, 18.

A glance at four decades of activity by the Jewish Agricultural Society (JAS) and its legacy for American Jewish farmers. Davidson compares the modern farm scene of 1940 with the struggle of farmers in 1900. In 1900 there were a few struggling Jewish farmers. As a result of the JAS efforts, in 1940 there are 100,000 Jewish families on farms in every state and every branch of agriculture. They live in modern homes, operate well stocked productive farms, and are active in farm and community affairs. The Society was a pioneer in credit unions, agricultural employment agencies and founded one of the first farm newspapers, *The Jewish Farmer*.

198. Davidson, Gabriel (1932 January 29). An Epic of the Prairies. *Detroit Jewish Chronicle*, 33 (10):1, 6–7.

An account of the establishment of the Cremieux Colony in the Dakota territory. The idea was conceived by Herman Rosenberg who had been a resident of the failed Sicily Island Colony in Louisiana. He thought that the temperate climate, good soil, and free land provided by the government were good incentives to try again. Davidson explains how Rosenberg gained support for the colony, how the site was chosen and the initial arrangements. The colony had about fifty settlers who were idealistic and hopeful that the experiment would work. The settlers had a good social life with parties, dances, lectures, meetings, and concerts but lacked farming experience and trained leadership. Cost overruns and debt, severe winters, intense summer heat that destroyed crops, and falling agricultural prices made it difficult to continue. Settlers began leaving and the colony was gone in three years. Davidson traced some of the followers to other cities where they entered new professions and businesses. (Also published in: Davidson, Gabriel. *Our Jewish Farmers and the Story of the Jewish Agricultural Society*. New York, L.B. Fischer, 1943, pp. 213–221.)

199. Davidson, Gabriel (1923 August 2). Jewish Farm Movement. Some of the Sociological Aspects Treated by the Manager of the Jewish Agricultural Society. *Ellenville Journal* (Ellenville, NY), pp. 1–2.

Davidson responds to skeptics who feel that farming is not a suitable livelihood for Jews by presenting the results of a survey sent to 608 randomly selected Jewish farmers in the East and Midwest. There were 470 respondents mainly foreign born who had been on their farms from 5 to 20 years. Questionnaires were sent to distinct communities that were representative of conditions throughout the country. An additional special survey was conducted among 138 farmers in a New Jersey district identified as the oldest Jewish agricultural community in the United States. Results indicated that nearly 86 percent were citizens or planned to become citizens. They considered themselves successful and content with their lives as farmers and happy to make America their home.

200. Davidson, Gabriel, and Edward A. Goodwin (1932 January 29). The Jewish Covered Wagon. *Jewish Criterion* (Pittsburgh, PA), p. 2.

The agricultural colonies of Moses Montefiore, Hebron, Gilead, Touro, Lesser, Beersheba, and Lasker were established in Kansas; all were small and short lived. Lasker was established in 1884 in the southwestern corner of Kansas forty miles from Ford City. It was named in honor of Eduard Lasker, a German economist and liberal thinker who died in New York City while touring. Davidson describes how the site was found, the plan of operation, social life, and hardships. Crop failures, low prices, physical deprivation, and isolation were too much for the settlers and the colony disintegrated within two years. The land was sold in 1886 and settlers left. (Also published in: Davidson, Gabriel. *Our Jewish Farmers and the Story of the Jewish Agricultural Society*. New York, L.B. Fischer, 1943, pp. 221–225.)

201. Davidson, Gabriel, and Edward A. Goodwin (1929 September 27). Chalutzim in the Land of Cotton. Down in Dixie an Attempt Was Made to Found Jewish Plantation Colony. *Jewish Tribune*, 95 (13):2, 15.

In 1881, a group of 125 Jewish immigrants consisting of tradesman, craftsmen, teachers, artists and other skilled professionals attempted to establish an agricultural colony in Louisiana. The colony, located at Sicily Island on the Ouachita River, was purchased with a grant of $8,000 from the Alliance Israelite

Universelle. Although the settlers were motivated by a pioneering spirit, the colony lasted about a year due a variety of factors which included poor land, crop failures, and a devastating flood from the Mississippi River that destroyed the settlement. The settlers abandoned the colony and left the area. Some joined other colonies that were being established in Arkansas, Dakota, and New Jersey. (Condensed version published as "Pioneers in the land of cotton" in: Davidson, Gabriel. *Our Jewish Farmers and the Story of the Jewish Agricultural Society*. New York, L.B. Fischer, 1943, pp. 204–208.)

202. Engel, Dave (1988 May 21). 40 Acres of Land and a Cow in Every Field. *Daily Tribune* (Wisconsin Rapids, WI), A3.

Brief description of the Jewish agricultural colony established at Arpin, Wisconsin in 1904. The colony began with 720 acres settled by twelve families who received forty acres each, a cow, two horses, a comfortable dwelling, and $5 per week for expenses. The Yiddish speaking settlement had twenty families at its peak. Max Leopold came to Arpin in 1908 at age fourteen when his father bought a forty acre farm; he later became a founder and member of Arpin. A student carrying out research on Arpin at the University of Wisconsin-Stevens Point interviewed Leopold in 1972 and provided anecdotes about the settlers and life at the Colony. Settlers observed the Sabbath and Jewish Holidays with services in private homes until 1916 when a synagogue was built. Men of the community took turns leading services and one of the settlers, Alvin Garber, served as Cantor and spiritual leader. Poor land, unfavorable weather, and economic hardship forced many settlers to leave and seek work elsewhere which led to the end of the colony by 1925. Leopold continued to farm until 1930. He was active in the Federal Land Bank, the Farm Credit Administration and the Wood County Farm Bureau. He ran for State Assembly (1920) and state Senate (1940) but lost both races. Leopold left in 1952 and settled in San Diego. Names of twenty former settlers are given.

203. Fishbein, Jennifer (2003 December 26). There Was a Chasid Who Had a Farm. Kosher Organic Farmers Find Mandate for Environmental Sustainability in Torah. *Forward*, 106(13477):12.

Describes Orthodox Jews who left their urban high pressure lives to establish farms in accordance with Jewish law. Shmuel Simenowitz, a Lubavitcher Rabbi, lawyer and Jewish educator, established Sweet Whisper Farms in Readsboro, VT where he raises organic produce and maple syrup. Yosef Abrams left a twenty six year law career and bought a 200 acre farm in Waukon, Iowa where he plans to create a kosher cheese business. Seven Lubavitch families founded a kosher organic farming community Eretz HaChaim in Sunderland, MA. They raise organic produce, cows and chickens according to Jewish law. Eretz

HaChaim is the only kosher farm involved in the Community Supported Agriculture program in which members buy shares in the farm and receive farm produce in return. These Chabad Lubavitch farmers live simply, enjoy the quality of farm life and claim that farming has a universal appeal that enhances the Jewish faith.

204. Geduld, Herb (2004 January 23). When Some of Our Forebears Were Farmers. *Cleveland Jewish News* (Cleveland, OH), 91 (4):A36.

From 1880 through 1910, groups of Jewish immigrants from Russia and Eastern Europe attempted to establish collective farms throughout the United States to rekindle a love of farming. They were assisted financially by the Alliance Israelite Universelle and the Baron De Hirsch Fund. Almost all failed due to internal bickering, lack of agricultural training and other factors. Four new Jersey settlements — Carmel, Alliance, Rosenhayn, and Woodbine — succeeded because they were well funded, highly motivated, and built factories to supplement the agricultural income. The first three settlements eventually became part of the city of Vineland, N.J., and lost their individual identities. Woodbine incorporated itself in 1903, became known as the "first self-governing Jewish community since the fall of Jerusalem." Geduld, whose uncle was a chicken farmer in Vineland, presents some historical background on the settlements.

205. Gruen, Oscar (1940 November 1). Jews in Agriculture. An Organized Farm Movement as a Constructive Program for American Jewry. Part I. *American Hebrew*, 147 (25):8, 14.

Gruen takes issue with those who claim that Jews cannot make a living in agriculture. He points out that Jewish involvement in agriculture goes back to biblical times. He describes an exhibit at a World Grain Conference he attended which had an exhibit arranged by the Jewish Colonization Association (JCA). The exhibit showed that there are Jewish farmers in ten countries and six continents. There are 100,000 farmers in the United States and the numbers are increasing annually. The exhibit displayed grain grown by Jewish farmers in Canada and Argentina and fruits, nuts, oils and wine produced by Jewish farmers in Palestine.

206. Gruen, Oscar (1940 November 8). Jews in Agriculture. An Organized Farm Movement as a Constructive Program for American Jewry. Part II. *American Hebrew*, 147 (26):10–11.

Gruen believes that a large proportion of Jews would become farmers if a well designed program were established to retrain Jewish youth in agriculture. He gives a few examples of individuals who have made important contributions to American agriculture. These include outstanding Jewish farmers, Nobel Prize winning scientists, founders of agricultural research institutes, professors of botany, plant physiology, and agricultural chemistry, and a poultry farmer who is renowned for his work on eradicating disease from poultry flocks. Jews have introduced important innovations in agriculture and organized and promoted vegetable and fruit industries and established worldwide markets. Agricultural settlements in Canada, Argentina, and Palestine have improved agricultural techniques in those countries. Gruen states that if a large scale effort were undertaken in America to get Jewish youth interested in agriculture, they will make great contributions in all areas of agriculture.

207. Haber, William (1940 November 1). The Refugee Farmer. The National Refugee Service Helps Farmers Back to the Soil. *American Hebrew*, 147 (25):8.

The National Refugee Service is helping refugees from Europe explore the possibilities of settling on farms. The National Refugee Service (NRS), in cooperation with the Jewish Agricultural Society, provides loans and offers training and consultation to refugees with no former agricultural experience. Since 1937, 196 families have been successfully settled on farms and NRS wants to expand the program by asking local Jewish communities throughout the country to assist with financial contributions to organizations involved in refugee settlement.

208. Hannah, Steve (1976 February 4). Jewish Settlers Found Land Was Cruel. *Milwaukee Journal*, p. 1, 5.

A nostalgic look back by former residents at the community of Jewish farmers at Arpin, Wisconsin established in 1904 with eighteen families. These residents who lived and worked at Arpin and knew the settlers and their families, offer their observations about why they came to Arpin, what the community offered them, their religious life and the synagogue, and the decline of the settlement after World War I. By 1922 only five of the original families were left and were still farming. The families began to leave when their children left the community to settle in other cities. The last resident left in 1952 and moved to San Diego.

209. Hartl, Harold (1936 August 6). State Worker Defends Colony. Denies Fraud, Communism. *Saginaw Daily News* (Saginaw, MI), p. 1.

Charles Sanders, Financial Officer of the Sunrise Cooperative Farm Community, testifies in Bay City Federal Court regarding a move to have a receiver appointed to investigate how funds received from the Rural Resettlement Administration were used. The money was supposed to be used to compensate members of the failing co-operative but claims of misuse were made. Divergent views about the Colony's financial status were given. Testimony by Sanders and other supporters refuted the claims and stated that no fraud occurred.

210. Hershenson, Sarah (2005 June 10). From the Heart: The Miracle Fund. A Lone, Frugal Farmer from New Jersey Left a Fortune in His Will to Help Make the Negev Bloom and Prosper. *Jerusalem Post*, Metro Israel Section, p. 10.

Before he died at age 96, New Jersey farmer Mack Ness donated his entire $20 million dollar estate to help Israeli entrepreneurs start businesses in Negev communities. Ness lived on a fifty acre farm near Watchung, NJ where he raised chickens, cows, and produce, and rented out a small number of homes on his farm. Although he never married and led a frugal existence, Ness was a skilled investor who amassed a multi-million dollar fortune. The money will be used to encourage agricultural ventures, hi-tech companies, new technologies, and small industries and stores throughout the Negev.

211. Hoffman, Mosa Heller (1944 September 15). Story of Cotopaxi. *Intermountain Jewish News*, 31 (37):16–17.

Compilation of articles about the Cotopaxi Colony in Colorado from *Denver Jewish News* (April 6 and 25, 1925) which became *Intermountain Jewish News* in (1926) and an article from the *Bulletin of the Colorado State Historical Society*. The articles present a brief history of Cotopaxi and personal interviews with four surviving members who tell their stories about the Colony.

212. Hoffman, David (2006 May 5). Strange Side of History: The First Jewish Town in the USA. *Yated Ne'Eman* (Monsey, NY), 18(17):19. Center section.

Biography of Baron Maurice de Hirsch (1831–1896), a philanthropist who devoted a major part of his fortune to establishing agricultural colonies in various countries to help Jewish immigrants fleeing the oppressive conditions in Czarist Russia. De Hirsch established the Jewish Colonization Association to facilitate settlement in Argentina, Canada, Australia and the United States. This article covers the Argentina and United States activities From 1892 to 1894. About 6,000 people were absorbed into Argentina, about half settled in the agricultural settlement and half remained in Buenos Aires and entered other occupations. The first successful United States colony was Woodbine, New Jersey, established in 1891 as a mixed agricultural/industrial settlement. By 1902 Woodbine had 1,400 people, fifty two farms and seven factories. Although it endured economic instability, the colony was largely successful until about 1920 when the community began to lose its largely Jewish character as non-Jews moved into the area. By 1945 there were only 545 Jews in Woodbine out of a total population of 1,792. and by 2003 Woodbine had a population of 2,716 which included only seventy two Jews. All that remains of the original settlement is the synagogue which is now a national historic site.

213. Imber, Naphtali Herz (1906 June 8). A Unique Jewish Colony—A Word to the Baron De Hirsch Fund. *Hebrew Standard*, 48 (21):1.

Garten Road is a farming community of about eighteen individual farm families not under the control of the Baron de Hirsch Fund located between the Jewish colonies of Rosenhayn and Carmel. Imber spent a vacation at Garten Road and was impressed by the independence of each family. The community had a library, two synagogues, a Hebrew school and a Jewish newspaper in each home. He believes it is a real agricultural community because it is devoted only to farming. Imber compares Garten Road to Carmel and Rosenhayn which in his view are poorly managed because they adhere to strict regulations imposed by the Hirsch Fund. He criticizes the colonies for straying from a pure agricultural orientation by adding small industries to supplement their income.

214. Itzkowitz, Shani (1999 July 9). Bought the Farm: A Jewish Family's Farm Encourages Torah Study and Respect for All Living Things. *Baltimore Jewish Times*, 248 (2):24.

Sara-Malka and Jacob Laderman moved from an apartment in Pikesville, Massachusetts to a nearby farmhouse. Baltimore-born Mrs. Laderman wanted a property large enough to create an organic vegetable garden; her husband wanted more space to work on cars. Over time they acquired goats, sheep, geese, doves, turkeys, a llama and a peacock and eventually established a vineyard. They call their farm "Jacob's Ladder" and provide farm tours and Torah study for guests and children.

215. Jacobson, Louis (2000 July 21). The Promised Land of Oz. *New York Jewish Press*, 79 (45):1.

A history of the Beersheba Colony which was settled in 1882 by sixty newly-arrived Jewish immigrants from Russia. Each family homesteaded 160 acres. Jacobson describes the settlement and the neighboring towns, land and soil conditions, and reasons why the colony lasted only two years. Includes a photo of a historical marker being placed on the site.

216. Johnson, Alvin S. (1940 November 1). Van Eden: A Jewish Farm Community. A Vivid Picture of an Experiment in Human Reconstruction. *American Hebrew*, 147 (25):5–6.

Van Eden is an agricultural settlement established in 1939 on a one thousand acre tract of land, thirty miles north of Wilmington, North Carolina. Alvin Johnson, Director of the New School for Social Research (New York), established a corporation to finance the settlement which is capable of holding seventy five families and suitable for market gardening. The present setting has nine immigrant families, mainly refugees from Nazi Germany, ranging in age from 35 to 50 on ten acres of land each and 150 acres in production. There are one or two children in each household. Johnson visited Van Eden to see how the settlers are getting along. He states that "all of them have grit." They are resourceful and independent, eager to learn, and support the settlement plan. They have been accepted by the community and other farmers and county agents visit regularly to offer advice and assistance. (For more on the Van Eden settlement see: Bentley, Melissa *The Van Eeden settlement: Alvin Johnson's attempt to rescue Jewish refugees and turn the tide of American public opinion in favor of Jewish resettlement in the 1930s and 1940s*. Senior Thesis. University of North Carolina at Asheville, 2003.)

217. Levitan, Tina (1951 October 5). Scientist Dreams of Life on a Farm. *Wisconsin Jewish Chronicle*, 68 (6):8.

A biography of Professor Albert Schatz, co-discoverer of Streptomycin. Schatz was born 1920 in Norwich, Connecticut. He developed a great love of nature and a keen interest in plants of all kinds during annual visits to his grandfather's farm in upstate Connecticut. By the age of four he began building terraria, and collecting fungi, mushrooms, mosses, lichens, rocks, minerals and other forms of plant life. He enrolled in the College of Agriculture at Rutgers and received his bachelors degree in soil chemistry and plant production in 1942. Schatz was interested in finding a drug to treat bacterial infections like tuberculosis. While working on a Ph.D. at Rutgers University which he received in 1945, Schatz and his associates discovered a new antibiotic agent produced by Actinomyces griseus which they called Streptomycin. He held various positions with research laboratories and was appointed Professor of Biology at Brooklyn college in 1949. He owns a 25 acre farm in Bergen County, New Jersey where he raises vegetables, lettuce, and celery. He plans to retire on the farm and carry out research on scientific farming and new methods of crop production.

218. Levy, Abraham R. (1905 November 24). The Jew as a Tiller of American Soil. *American Hebrew*, 77(26):849–864.

In this 3 part article, Levy discusses the need to settle poor immigrants from Russia, Roumania and Galicia in agricultural communities. In Part I, "Immigration and Farming," he states that the crowded industrialized cities can not be relied upon to absorb the large number of new immigrants and provide for their well being. He proposes that they be encouraged to make agriculture their way of life and shows how the Jewish Agriculturists Aid Society of America is involved in settling the immigrants on farms throughout the country and meeting their financial and housing needs with loans and grants. Levy extolls the virtues of agricultural life and asserts that it makes the immigrant an independent and self reliant human being. He gives brief examples of some of the successful Jewish farm families in Michigan and North Dakota. Part 2, "The Jewish Town of Woodbine" by Kellogg Durand from the *Boston Transcript*, is a study of the unique agricultural settlement at Woodbine, New Jersey and its philosophy, activities, farming methods and the National Farm School. Part 3, "The Arpin Colony" from the *Evening Wisconsin*, contains the observations of a staff correspondent who visited the settlement at Arpin, Wisconsin. He explains how the colony was started, the settlement plan, the land, work ethics, and daily life.

219. Levy, Abraham R. (1901 January 25). The Scope and Aim of the Movement to Promote Husbandry Among the Jewish Poor. *American Hebrew*, 68 (10):317–318.

Levy points out that the dedication of the Baron de Hirsch Agricultural School at Woodbine in November 1900 offers the opportunity to train Russian Jews in this country to take up farming. He states that farming is a noble occupation for Jews, but the Jew of his day is unfamiliar with farming and particularly the Russians who were denied the opportunity to enter agricultural professions. Levy believes that if a Jew is mentally and physically fit and has a desire to farm, it is attainable. If the Russians take the initiative to apply themselves and learn the rudiments of farming, the school can provide the appropriate education for a successful and contented life as farmers.

220. Levy, Abraham R. (1901). The Promotion of Husbandry Among the Jewish Poor: Past Achievements — Present Needs. *American Israelite*, 47 (50):1.

A description of the activities of the Jewish Agriculturists Aid Society in helping poor immigrants get established in agricultural occupations. Using Jewish farmers in North Dakota, Michigan and Illinois as examples, Levy explains the financial and technical assistance that the Society provides and the number of farmers who have become self sufficient and free of debt. He indicates that the number of individuals seeking assistance has grown and makes a plea for more donors to come forward and help the Society settle more farmers and assist in the purchase of needed machinery and equipment.

221. Lipman, Steve (2005 February 25). Greenberg Acres: Take That, Larry Summers: American Jews Are Fueling a Farming Revival. *New York Jewish Week*, 217 (40):1, 25.

A response to Lawrence Summers, President of Harvard University, who commented that there is a paucity of Jews in farming in the United States. Lipman points out that although Jewish farmers comprise a small percent of the Jewish population, there are Jewish farmers in every state raising cotton, poultry, cattle, or whatever is appropriate to their location. The numbers are growing with younger, better educated, mostly Orthodox turning to farming. Some enter farming in later years to fill a desire to farm as a second career. Lipman gives a brief history of Jewish farmers in early America and brief sketches of some of today's Jewish farmers. These include Benjamin Cantz, who operates a winery in Santa Cruz, CA; Rabbi Shmuel Simenovitz, who produces maple syrup in Vermont; David Jakobs, who operates a family farm in Northern Illinois; Mike Tabor, a truck farmer in Needmore, PA; Eretz Chaim, a fish farm started by Lubavitch Jews in western Massachusetts, and Dan Glickman, a nine term congressman from Kansas who served as Secretary of Agriculture under President Clinton.

222. Mark, Jonathan (2004 June 18). The Country Life. The Jewish Farms Failed, But There's Nothing to Regret. Come, Then Down to Southern Jersey, and Back to a Sweeter Time. *Jewish Week* (Celebrating 350, 1654–2000: The Road to Jewish America — A Jewish Week Commemorative Issue), pp. 20–23.

A short history of Jewish farming in the Catskills, Connecticut, and New Jersey. Mark focuses on the chicken and egg farmers in Vineland and Woodbine in Cumberland County, New Jersey that existed from 1880 through 1920. Some remained in the area as independent farmers through the 1970s when competition from large scale farming forced them out of business. Today in Cumberland County, there are 1,285 Jews with only a handful still in the chicken business. Includes anecdotes from farmers and family members who grew up on the chicken farms about their experience working with chickens, hard work, cold winters, their religious practices and synagogue attendance, and other nostalgic reminiscences of a bygone era in New Jersey agriculture.

223. Newman, Andy (2005 June 22). The Last of the Jewish Farm Girls. Nearing 100, a Survivor Recalls a New Jersey Colony. *New York Times*, Metro Section, pp. B1, B6.

Lillian Greenblatt, the last surviving resident of the Jewish agricultural colony of Alliance, New Jersey returns to visit the Colony's cemetery as she nears her 100th birthday in August, 2005. Greenblatt grew up in a section of Alliance known as Brotmanville named for Abraham Brotman, who built a clothing factory where settlers earned additional income during the winter months. The Alliance colony, established in 1882, had a population of 500 at its peak and was active enough to support four synagogues. Greenblatt's father came to Alliance in 1890 and settled in Brotmanville where he got a job with a local builder and built houses for the Brotmanville residents. He later became a kosher butcher. By the mid 1920s only a handful of Jews were left in Alliance. Greenblatt moved with her parents to Vineland in 1920 where she married and ran a children's clothing store. She is widowed and now lives in Maryland. She tells anecdotes about her childhood and life in the colony. Today, Brotmanville is mainly African American and one of the early synagogues is a Baptist church. Another synagogue across the street from the cemetery still stands and is used for high holiday services by Jews who want to preserve its history. Includes a brief comparison with the Woodbine colony which grew to 2,000 people and became a separate municipality in 1903.

224. Palitz, Bernard A. (1909 October 1). Jewish Charity vs. Jewish Agriculture. *Jewish Exponent,* 49 (25):1–2.

Palitz issues a challenge to the Jewish charities to widen their focus on helping Jews by including support for agricultural programs. He states that about 5000 Jewish farmers in the country are doing well and find farming a profitable and resourceful occupation. The Jewish Agricultural and Industrial Aid Society has established the framework for settling immigrants on farms. Palitz asserts that it is now incumbent upon the Jewish Charities that have usually focused on assisting urban Jews, to come together and participate in the agricultural programs and help Russian immigrants adapt to life on the land.

225. Rich, Adolph W. (1951 September 28). Wisconsin's Only Jewish Colony Begun at Arpin, 1904. The Story of the Founding and Progress of a Small Agricultural Settlement. *Wisconsin Jewish Chronicle,* 68 (5): 12–13.

Text of a paper read at the National Conference of Jewish Charities, Philadelphia, May 8, 1906, telling the story of the Jewish Agricultural Colony at Arpin, Wisconsin. Adolph W. Rich, a Milwaukee merchant, manufacturer, and philanthropist, explains how he was instrumental in organizing the colony with the assistance of the Milwaukee Agricultural Society and the Jewish Agricultural and Industrial Aid Society in 1904. The colony comprised 1000 acres and held 17 families. Rich discusses how the colony was established, land clearing, making roads, building houses, digging wells, and other jobs to prepare for working and living at Arpin. Each family received one or two cows, a horse, a wagon, and necessary implements, and an income of $5 per week. Although Rich believed that the colonists were content and looked forward favorably to life at Arpin, discouragement, dissatisfaction, family troubles, cold winters, and bad soil conditions, led to the eventual breakup of the colony. All that stands at Arpin today is the remains of a synagogue built about 1915. (Also published in: *Proceedings of the Fourth Biennial Session of the National Conference of Jewish Charities in the United States.* May 6–8, 1906, pp. 170–181.)

226. Rudman, H. David (1973 January 4, 11, 18, 25, February 1). The Sholem Colony the First American Jewish Rural Settlement, 1837–41. *Ellenville Journal* (Ellenville, NY), various pages.

The Sholem Colony in Wawarsing, NY began in 1838 with twelve families who contracted for the erection of houses, a synagogue, a store, two factories, and a museum and art gallery. The settlers were influenced by the back to the land movement of the time and were determined to succeed. The land was barren, rocky and more than 100 miles from New York City. Economic difficulties, the mountainous terrain, a short growing season, lack of farming experience and differences regarding operation of the colony were prevalent from the outset. The colony lasted four years and was abandoned in 1841 following foreclosure. In this five part article, Rudman provides details about land acquisition and mortgage arrangements, and names and personal information about the settlers and their backgrounds. He analyzes the difficulties they faced, reasons for the failure of the colony, and comparisons with other failed settlement projects of that era.

227. Saltiel, Emanuel H. (1882 October 27). Mr. Saltiel and the Cotopaxi Colony. *American Hebrew*, 12 (11):132.

A letter from Emanuel Saltiel responding to allegations by Morris Tuska of the Hebrew Emigrant Aid Society (HEAS) that he misused funds given to him by the Society to assist the settlers at Cotopaxi colony. In this point by point rebuttal, Saltiel explains his conflicts with Morris Schwarz, who was assigned to oversee the colony but had different ideas about how to allocate the funds. Saltiel claims that Schwarz used superstition and bigotry to turn the settlers against him which made it difficult to meet his obligations. He explains his side of the story and his relationship with the settlers and HEAS which was favorable at the early days of the colony but became hostile following Schwarz failed schemes. (See: Tuska, Morris. The Cotopaxi Colony. *American Hebrew* 12(8):92, October 4, 1882.)

228. Schachter, Harry A. (1987 October 4). A Kingdom Was Built of Chickens and Chores. *New York Times*, p. WC46.

Reminiscences of life on a 60 acre poultry farm near Croton Falls in Putnam County, New York. Schachter purchased the farm in 1953 and moved there with his wife and three children. They raised 10,000 birds that produced about 4,000 eggs per day. He sold the farm in 1967 and the house in 1969 and moved to Croton-on-Hudson.

229. Stringer, William H. (1943 May 8). The Jew with a Hoe. *Christian Science Monitor,* Weekly Magazine Section, pp. 6, 13.

A tribute to the 100,000 Jewish farmers around the country who are cultivating nearly 1,500,000 acres on holdings estimated at more than $150,000,000. Since 1937 refugees from Germany, Poland, Holland, Belgium, France and some from crowded American cities have found new lives in agriculture with the aid of the Jewish Agricultural Society. Stringer offers short stories about some of the farmers who have been "rehabilitated and restored in a new environment." Stringer predicts a bright future for agriculture and recommends that Jewish and non-Jewish leaders be more forthcoming with funds and publicity to encourage more Jews to engage in farming.

230. Tuska, Morris (1982 October 4). The Cotopaxi Colony. *American Hebrew,* 12 (8):92.

A report by the Executive Director of the Hebrew Emigrant Aid Society regarding conditions at the Cotopaxi Colony. Focusing on the poor financial conditions of the colony, Tuska attributes the situation to Emanuel Saltiel, who offered to provide housing and equipment for the colonists and help them to become self sufficient. Saltiel was given funds by the Hebrew Emigrant Aid Society to assist in the settlement effort. Tuska alleges that Saltiel used the money for his own purposes and left bills unpaid. Tuska put the management of the Colony in the hands of a settler named Julius Schwarz. HEAS will work with Schwarz to help ease the economic situation of the settlers. (See Saltiel's response to Tuska in: Saltiel, Emanuel. Mr. Saltiel and the Cotopaxi Colony. *The American Hebrew,* 12(11):132, October 27, 1882.)

231. Warbasse, J. P. (1930 March 14). Why Be a Farmer. *Mountain Hotelman* (Ellenville, NY), 1 (1):1–2.

Warbasse laments the tendency of intelligent rural youth to move to the cities. This has resulted in a process of natural selection that led to the backwardness of agricultural communities. He claims that the back to the land movement will attract superior individuals who are suited for farm life and the peace and solitude of country living. These individuals enjoy the variety of interests and talents that farming offers and allows a rich relationship with neighbors. In Warbasse's view, the farmer occupies a strategic position in times of turmoil and depression because he can protect his family by raising food on his own land.

232. Wartzman, Rick (1999 April 27). Exacting Practice Makes Near-Perfect for California Esrogs — The Fruit Is Pricey, and Spots Aren't Kosher; the Rules Are a Tough Row to Hoe. *Wall Street Journal* (Eastern Edition), pp. A1, 8.

Story of John Kirkpatrick, a 68 year old farmer who is the only large scale U.S. producer of the esrog, known in English as a *citron*, a yellow fruit used to celebrate the Jewish holiday of Sukkot. Kirkpatrick, a Presbyterian, grows mostly lemons and tangelos on thirty seven acres in Tulare County, California, in the San Joaquin Valley. He has 253 esrog trees which get special attention to meet the specifications of Jewish consumers. Avrohom Teichman, a Los Angeles Rabbi, is responsible for certifying that the esrogs are grown according to Jewish law. Kirkpatrick's trees can sprout about 100,000 esrogs, but only about 1,200 are cultivated to full size. Yaakov Rothberg, his business associate, is an Orthodox Jew who comes from his home in Lakewood, NJ, to grade the fruit and choose about 2,000–4,000 fruits that are good enough to ship to the New York area markets.

233. Wiener, Julie (2004 May 28). The Greening of Ansche Chesed. *New York Jewish Week*, 216 (52):1.

The kosher food industry is responding to the growing interest in eco-agriculture and the rising demands for organic kosher foods. Congregation Ansche Chesed, a conservative synagogue on the Upper West Side of Manhattan, formed a partnership with a Jewish farmer on Long Island to supply organic food to its cooperative. Kosherfest, the annual kosher foods trade show, is expanding its offerings of organic foods to meet demands. Two North American businesses now distribute kosher free-range organic poultry. Wise Kosher Poultry, based in Williamsburg, Brooklyn, run by a Satmar Hassidic mother of eleven children, is planning to offer kosher free-range organic beef as well. An organic farm offers summer work and learning under auspices of the Isabella Freedman Jewish Retreat Center. Eretz Ha'Chaim, a Torah-observant organic farm cooperative near Amherst Massachusetts, sponsors an annual festival and plans to create an educational center for children and adults.

234. Wiener, Julie (2004 December 27). Judaism's New Fertile Soil. In the Small But Growing Environmental Movement, Young Jews are Getting Physical and Going Back to the Land. *New York Jewish Week, Directions Magazine (Supplement to the Jewish Week)* 217(32):18–23.

A summer program combining organic farming and Jewish studies at the Adamah Retreat Center in the Connecticut Berkshires offers students a chance to combine environmental values and stewardship of the earth with Judaism. The program attracts college and pre-professional students who work on the 4 acre farm, participate in bikeathons with other Jewish youth, perform chores, eat in the communal cafeteria and pray, meditate, and learn. Programs focusing on Jewish teenagers are also being planned. The Center's goal is to connect Jewish spirituality and environmental values and to demonstrate to Jewish youth that concern about the environment is a part of Jewish tradition.

235. Wilensky, Melody (2000 January 27). On Being a Jewish Farmer: An Interview with Michael Doctor. *Jewish Advocate*, 190 (3):1.

Michael Doctor started the Food Bank Farm in 1990 as the Chili Project, a joint effort of the Food Bank of Western Massachusetts and Hampshire College, to provide chili and vegetables for soup kitchens and shelters. The Farm has over 510 shareholders representing 900 families and grows more than 40 different types of organic vegetables. The produce is also distributed to the agencies served by the Food Bank of Western Massachusetts which supplies nearly five million pounds of food annually to 400 local social service programs. Doctor points out that these programs reach more than 100,000 needy people throughout four counties of western Massachusetts. Doctor believes that this program allows members to connect to Jewish values through land and food. He discusses his Jewish background and education and how he became interested in farming.

Books

236. Adler, Cyrus (1928). *Jacob H. Schiff: His Life and Letters.* Garden City, NY: Doubleday, Doran. 2 vols. 794 pp.

A biography of Jacob H. Schiff (1847–1920), banker, philanthropist, statesman and a leader in the American Jewish Community. This two volume work based upon letters, correspondence and other primary sources presents Schiff's life and his influence on matters relating to Jewish issues. Includes coverage of his relationships with the Baron Maurice De Hirsch and the agricultural colonization movement. Schiff participated in the activities of the Hirsch Fund but expressed his concern with some economic aspects of settlement programs.

237. Allman, Herbert D. (1935). *A Unique Institution: The Story of the National Farm School.* Philadelphia: Jewish Publication Society of America. 238 pp.

The National Farm School was established on April 10, 1896, near Doylestown, Pennsylvania, 26 miles from Philadelphia. The purpose of the school was to instruct Jewish youth in the theory and practice of agriculture. The book contains two parts: Part I covers the formation, development, expansion and progress of the school including facts about the instructors, students, educational programs, economics, vocational guidance, alumni affairs. Part II summarizes the organization and functions of the various departments and activities in the dairies, poultry plants, greenhouses, laboratories, and field work. The appendix includes lists of guests, endorsers, directors, teachers, and staff who supported the school and its programs.

238. Aron, Bill, Vicki Reikes Fox and Marie Cohen Ferris (2002). *Shalom Y'all: Images of Jewish Life in the American South.* Chapel Hill, NC: Algonquin Books of Chapel Hill. 159 pp. ISBN 1565123557.

Photographs and stories about Jewish life in the South from the Museum of the Southern Jewish Experience. Depicts Jews in all walks of life at home, business, and synagogue. Includes a cotton, wheat and peanut grower in the Mississippi Delta, cotton merchants in Montgomery, Alabama, and a pecan shelling operation in Natchez, Mississippi.

239. Ashkenazi, Elliott (1988). *The Business of Jews in Louisiana, 1840–1975.* Tuscaloosa: University of Alabama Press. 231 pp. ISBN 0817303189.

This work based upon the author's PhD dissertation *Creoles of Jerusalem: Jewish Businessmen in Louisiana, 1840–1875*, explores the role of Jews as businessmen, merchants, farmers, and plantation owners in Louisiana from 1840–1975. Ashkenazi examines the commercial activities of Jews within the State, prominent Jewish families in agriculture, and the importance of sugar and cotton in the economy of that period. (See also: Ahskenazi, Elliott [1983]. *Creoles of Jerusalem, Jewish Businessmen in Louisiana, 1840–1875.* PhD dissertation, George Washington University.)

240. Berman, Hyman, and Linda Mack Schloff (2002). *Jews in Minnesota.* St. Paul, MN: Minnesota Historical Society. 102 pp. ISBN 0873514181.

Jews have made important contributions to the business, politics, and education in Minnesota since its early days. In 1891, New York investment banker Jacob H. Schiff leased land from the Great Northern Railway to establish an agricultural settlement at Milaca in Mille Lacs County. Arrangements were made to build about fifty houses on forty acre plots to house the settlers who were given about $600 to get started. By 1898 only thirteen families had been settled but the fate of the colony is unknown.

241. Blood, William W. (1973). *Apostle of Reason: A Biography of Joseph Krauskopf.* Philadelphia: Dorrance & Company. 271 pp. ISBN 0805918612.

Biography of Joseph Krauskopf (1858–1923), reform rabbi, philosopher and social reformer. Krauskopf supported the efforts of Baron Maurice de Hirsch and the Jewish Agricultural Society to establish agricultural settlements for Russian immigrants. Krauskopf believed that the settlements will succeed if the settlers and their families had access to the latest information on agricultural techniques. He founded the National Farm School (now Delaware Valley College) and served as its Director. This biography covers his life and his contributions to Reform Judaism and agricultural education.

242. Brandes, Joseph, and Douglas Martin (1971). *Immigrants to Freedom: Jewish Communities in Rural New Jersey since 1882.* Philadelphia: University of Pennsylvania Press. 437 pp. ISBN 0812276205.

A comprehensive history of the establishment and development of the Jewish agricultural colonies of Rosenhayn, Carmel, Garton Road, Norma, Alliance, and Woodbine in Southern New Jersey. This detailed work takes readers on a journey from the arrival of immigrants in the 1880s and their orientation to rural life in America through the establishment, development, progress, and decline of the colonies. Recounts the challenges they faced and how they handled difference of opinions and conflicts. The Appendix contains a roster settlers of the Alliance colony and statistics on the population, social life, industrial conditions, religious life, public school education, and economic conditions of the six colonies in 1919; the New Jersey egg industry, 1940–1964, and New Jersey agriculture, 1950–1968.

243. Breck, Allen DuPont (1960). *The Centennial History of the Jews of Colorado, 1859–1959.* Denver: Hirschfeld Press. 375 pp.

A one hundred year history of Jewish settlement in Colorado and the contributions of Jews in all aspects of Colorado life and development including farming and cattle raising and the history of the short lived Cotopaxi and Atwood agricultural colonies. Provides facts about the founding of the colonies, support groups, names of settler families, and the conflicts and adverse conditions that contributed to the failure of the colonies.

244. Cahan, Abraham (1969). *The Education of Abraham Cahan.* Philadelphia: Jewish Publication Society of America. 468 pp.

Cahan was a founder and editor of the *Jewish Daily Forward.* In this autobiography, he presents the story of his life from his birth in a Vilna, Lithuania in 1860 to his arrival in New York in 1882 and the start of his journalistic career in 1890. Cahan was a member of Am Olam and still considered himself a socialist during his early years in America. He knew some of the individuals involved in the Oregon and New Jersey agricultural colonies and the back to the land movement. In various places in the book he offers his observations about the colonization movements, the people he knew, and why he felt the colonies would not survive. (Translated by Leon Stein, Abraham P. Conan, and Lynn Davison from the Yiddish autobiography *Bleter Fun Mein Leben.*)

245. Calof, Maier (1941). *Miracles of the Lives of Maier and Doba Calof.* Privately published. 86 pp.

This is a story of the life of Maier Calof who was born in Steblev, near Kiev in 1868 and emigrated to the United States at age 15 to join his father Mordechai who emigrated a short time earlier and settled in the Painted Woods Colony in the Dakota Territory. Other relatives arrived later including his cousin Doba whom he married. The Calofs left Painted Woods after thirteen years and moved to Winnipeg where Maier operated a meat business, a timber business and wholesale merchandise enterprise. They moved to Los Angeles in 1944 and became United States citizens. The Calofs were married fifty three years and raised ten children. In this stirring narrative, written at age seventy six, Maier Calof talks about his life in Steblev, his immigration to America, the harsh realities of daily life in Painted Woods. He recounts their experiences in Canada and Los Angeles, anecdotes about their children and the contrast between those times and the Steblev and Painted Woods years. Maier was an observant Jew who quotes scripture and Jewish history and tradition throughout the book. (See also: Calof, Rachel B. and J. Sanford Rikoon [1941]. *Rachel Calof's Story: Jewish Homesteader on the Northern Plains.* Bloomington: Indiana University Press.)

246. Calof, Rachel B., and Sanford J. Rikoon (1995). *Rachel Calof's Story: Jewish Homesteader on the Northern Plains.* Bloomington: Indiana University Press. 171 pp. ISBN 0253329426.

This four part memoir and history contains "My story" by Rachel Bella Calof; "Epilogue" by Jacob Calof; "Jewish farm settlements in America's heartland" by J. Sanford Rikoon, and "Rachel Bella Calof's life as collective history" by Elizabeth Jamieson. Calof (nee Kahn), born in Russia in 1876, tells the story of her deprived childhood in a household with a cruel stepmother and weak father and the hardships that she underwent until leaving Russia in 1894. She migrated to the United States for an arranged marriage with Abraham Calof and lived with her husband and his family as homesteaders near Devils Lake in northeast North Dakota from 1894–1917. In this memoir, translated from Yiddish, Calof discusses their lives as homesteaders and the persistent poverty, harsh weather, crop and animal losses, lack of privacy, and illness. Rachel Calof bore nine children, took care of her home and did hard work on the farm with a determination and courage that helped the family survive. Her household also served as a meeting place for other Jewish farmers in the region. The Calofs left North Dakota in 1917 and settled in St. Paul. Jacob Calof, Rachel's son, offers some observations about his mother's strength of character and her endurance as a Jewish farmer's wife on the Dakota Prairie. Rikoon analyzes the plight of Jewish settlers in the Heartland between 1880 and 1840. Jamieson interprets class, culture, place, and gender in the rural Midwest during those years and Rachel's determination to express her Jewishness in that environment.

247. Cohen, David (1989). *The Jews in America.* San Francisco: Collins Publishers. 223 pp. ISBN 0002153238.

Contains two hundred captioned photos showing a photographic cross section of the American Jewish community and the religious and secular aspects of their daily lives. Includes photos of Scott Gerber, who runs a cattle ranch in Petaluma, California, and Norbert Jakobs who raises 6000 head of beef cattle on his farm in Sterling, Illinois (pp. 118–121).

248. Cohen, Joseph J. (1957). *In Quest of Heaven: The Story of the Sunrise Co-Operative Farm Community.* New York: Sunrise History Publication Committee. 271 pp. (Reissued: Philadelphia, Porcupine Press, 1975. ISBN 0879910232.)

The Sunrise Co-operative Farm Community, a libertarian collectivist colony in Saginaw Valley, Michigan, existed from 1933 to the spring of 1938. It was one of many experiments in communal living that arose in response to the hardships of the depression era. This is a detailed account of the establishment, development and dissolution of the colony as told by its founder. Cohen explains the influence of collectivist ideologies that motivated the members. Members had not anticipated the ever present strains that hard labor, debt, weather, crop failures, lack of agricultural training, and economic hardships have on the collective nature of their existence. Differing personalities and political attitudes added an additional strain that caused many heated debates and conflicts among members. Cohen reflects on the good and bad years at Sunrise and offers some insights about human nature that led to the fading of the dream and the end of the experiment.

249. Cohen, Naomi W. (1984). *Encounter with Emancipation: The German Jews in the United States, 1830–1914.* Jewish Publication Society: Philadelphia. 422 pp. ISBN 0827602367.

A study of the American experience of German Jewish immigrants who came to the United States during the 19th and early 20th centuries. This book examines the challenges they faced, the choices they made, and their adaptation and integration into American life. Among the many areas of involvement that were open to the immigrants, agriculture was not of great interest to the German Jews. Various attempts were made to encourage them to settle in agricultural colonies but they were short lived and most of the settlement programs geared to German Jews were discontinued. Cohen briefly covers the settlement movements and explains why they were not widely accepted.

250. Cohen, Naomi W. (1999). *Jacob H. Schiff: A Study in American Jewish Leadership.* Hanover, NH: Brandeis University Press. 334 pp. ISBN 0874519489.

Jacob H. Schiff (1847–1920) was a banker, financier, philanthropist and prominent Jewish leader. He was a spokesman for the Jewish community and was active in helping Jewish immigrants get acclimated to American life. This extensive biography documents Schiff's activism on behalf of American Jews. Describes his relationship with Baron Maurice de Hirsch and the efforts to help immigrants from Eastern Europe settle on farms and pursue a life in agriculture.

251. Cowen, Philip (1932). *Memories of an American Jew.* New York: International Press. 438 pp. (Reprinted: New York, Arno Press, 1975. ISBN 0405067038.)

Philip Cowen (1853–1943), was the founder and editor of the weekly newspaper *The American Hebrew* from 1879 until 1906. He was active in Jewish affairs and wrote about Jewish life in America. In this autobiography, he offers his views on immigration, life in the cities, Jewish life in America, Jewish life in Russia and Roumania, and interesting Jewish personalities of the time. Contains scattered coverage of the agricultural settlement efforts and the activities of the Baron De Hirsch Fund and the Jewish Agricultural Society. Includes a letter (p. 100) from Bennie Greenberg, a North Dakota farmer, explaining how he makes a living as a farmer and advising Jewish farmers not to accept financial assistance. Greenberg states that his success as a farmer was due to his own initiative and not dependent upon help from others.

252. Daly, Charles P., and Max J. Kohler (1893). *The Settlement of the Jews in North America.* New York: Phillip Cowen. 189 pp.

This is an edited and expanded work based upon a speech Daly presented to the Hebrew Benevolent Aid Society on April 11, 1872. It presents a history of Jewish settlement and involvement in American life from the time of Columbus through 1893. Focuses on New York, Pennsylvania, Maryland, Georgia, South Carolina and Rhode Island, Delaware, the American colonies, and sketches of prominent New York Jews. References to agriculture include W.D. Robinson's London "Memoir" (1819) addressed to wealthy Jews encouraging them to purchase large tracts of land in Mississippi and Missouri territory for agricultural settlements and Mordecai M. Noah's plans for a Jewish colony, Ararat, at Grand Island, New York. (See also: Robinson, William David. *Memoir addressed to persons of the Jewish religion in Europe on the subject of emigration to, and settlement in, one of the most eligible parts of the United States of North America.* London, H. Hay, 1819.)

253. Davidson, Gabriel (1943). *Our Jewish Farmers and the Story of the Jewish Agricultural Society.* New York: L.B. Fischer, Printed and bound by Kingsport Press, Kingsport, TN. 289 pp.

The Jewish Agricultural Society was founded in 1900 to assist immigrant Jews from Eastern Europe to settle on farms and gain their livelihoods in American agriculture. This book presents the Society's history, activities, and accomplishments over a forty year period — the formative years; the period of exploration and experimentation; the period of expansion and evaluation, and the depression years which forced a reshaping of the society due to dwindling funds. The supplement (pp. 195–275) contains reprints of articles from Jewish periodicals on the various agricultural colonies, a history of the National Farm School, and the names of directors and staff of the Jewish Agricultural Society.

254. Dobin, Abraham (1975). *Fertile Fields: Recollections and Reflections of a Busy Life.* South Brunswick, NJ: A.S. Barnes. 519 pp. ISBN 0498015459.

Abe Dobin grew up on a farm in South Brunswick, New Jersey and spent his life serving New Jersey agriculture and his community. In this autobiography, spanning a sixty five year period, he chronicles his life on the farm; eighteen years as an agricultural extension agent with the Jewish Agricultural Society; his business ventures; founding of a farmers cooperative in Hightstown, New Jersey; organizer of a 4-H club; involvement in building the first synagogue in New Brunswick, and political years as a Committeeman, member of the zoning board and two time Mayor of South Brunswick. This work offers a panoramic view of Dobin's remarkable life and his family, friends, and associates.

255. Dubrovsky, Gertrude W. (1992). *The Land Was Theirs: Jewish Farmers in the Garden State.* Tuscaloosa, AL: University of Alabama Press. 251 pp. ISBN 0817305440.

Dubrovsky, who grew up on a poultry farm in Farmingdale, New Jersey, presents a fascinating look at a unique community that helped make New Jersey one of the leading egg producing regions in the United States. Using data from personal interviews with 120 former residents, personal experience, and life histories, she recounts the lives of Jewish farmers and their families in the Farmingdale area from the beginning of settlement in 1919 through the 1950s. The book abounds with anecdotes about the farms, the families who worked the land, relationships with neighbors, education, Jewish life and institutions, and the triumphs and struggles during three decades of rural life in Farmingdale.

256. Eisenberg, Ellen M. (1995). *Jewish Agricultural Colonies in New Jersey, 1882–1920.* Syracuse, NY: Syracuse University Press. 241 pp. ISBN 0815626525.

A revised edition of the author's PhD Dissertation *Jewish agricultural colonies in southern New Jersey: the processes of migration, settlement and adaptation* (1990). Eisenberg examines the patterns of migration and adaptation and the social, religious, and cultural aspects of Russian immigrants in the agricultural colonies of Salem and Cumberland Counties, New Jersey from 1880 through 1920. Focusing on Alliance, Norma and Brotmanville in Salem County and Rosenhayn and Carmel in Cumberland County, she analyzes characteristics and ideologies of the settlers, the role of sponsoring organizations, the early years of settlement (1882–1890), transformations from a communal to a mixed agricultural/industrial colony based on private ownership (1900–1910), and dissolution of the colonies in their final years (1911–1920).

257. Feldberg, Michael (2005). *Three Hundred and Fifty Years: An Album of American Jewish Memory.* New York and Cincinnati: American Jewish Historical Society; Jacob Rader Marcus Center of the American Jewish Archives. 258 pp.

This album contains photographs and text depicting Jews in all areas of American life during the 350 years since the first Jewish settler arrived. Jewish farmers are covered in a brief history (p. 114) and photograph of farm products from New Jersey about 1910 (p. 115). ("Published on the occasion of the National Dinner by "Celebrate 350: Jewish Life in America," in cooperation with The Commission for Commemorating 350 Years of American Jewish History, the American Jewish Historical Society, the Jacob Rader Marcus Center of the American Jewish Archives, the Library of Congress, and the National Archives and Records Administration.")

258. Fogarty, Robert S. (1990). *All Things New: American Communes and Utopian Movements, 1860–1914.* New York: Cambridge University Press. 293 pp. ISBN 0226256545.

The period from the 1860's to early twentieth century was a time when groups of people who shared common interests in communal life styles were attempting to form cooperative living arrangements. Fogarty describes these movements and the philosophical and cultural factors that motivated them. In referring to immigration by Russian Jews (pp. 112–117), he points out that various schemes were in place as far back as 1783 to encourage Jews to emigrate to the United States and other western countries to settle in agricultural colonies. Fogarty points out that the Am Olam movement which was founded in Russia was the incentive for establishment of twenty six colonies in eight states. Fogarty describes some of those projects and the differing philosophies and factors that led to their successes or failures.

259. Glazier, Jack (1998). *Dispersing the Ghetto: The Relocation of Jewish Immigrants Across America.* Ithaca: Cornell University Press. 255 pp. ISBN 0801435226.

The early waves of large Jewish immigration from Russia and other East European countries began in the 1870s and continued through the 1920s. The Industrial Removal Office was established by the Jewish Agricultural and Industrial Aid Society (later the Jewish Agricultural Society) in 1901 to assist in dispersing the new arrivals to urban and rural communities around the country. For two decades, IRO resettled over 80,000 people to over 1000 cities and towns until its dissolution in 1922. This book is an analysis of IRO's activities and challenges over a twenty year period. Covers the distribution procedures, immigration debates, immigrants' views regarding their adaptation to America, operation of the IRO in several Midwest communities, and comparison of contemporary immigration issues with the earlier era. Includes the Jewish agricultural colonies and the IROs role in dispersal and settlement and its impact on the lives of the immigrants.

260. Goldberg, Robert Alan (1986). *Back to the Soil: The Jewish Farmers of Clarion, Utah and Their World.* Salt Lake City: University of Utah Press. 224 pp. ISBN 0874802636.

In 1911, two hundred immigrant Jewish families left the urban areas of New York, Philadelphia and Boston to begin life on the land as farmers in a settlement in Clarion, Utah with financial assistance and support from the Jewish Agricultural Society. The colony lasted five years, but a dozen families remained on the land until the ultimate demise of the colony in the mid-1920s. Goldberg, a Professor at the University of Utah, documents the history of the Clarion settlement and the people, places, and community structure from interviews with former Clarion residents, archival sources, and newspapers. He describes the politics, poor land and soil conditions, hardships, lack of agricultural training, inadequate finances, and the threat of assimilation that led to the end of the Clarion settlement.

261. Hamburg, Alice, Tova Gazit and Ruth Kelson Rafael (1989). *Sam Hamburg, Agricultural Pioneer in California and Israel: A Biography by Alice Hamburg with a Guide to the Hamburg Archive at the Western Jewish History Center.* Berkeley, CA: Western Jewish History Center, Judah L. Magnes Museum. 87 pp. ISBN 0943376424.

Sam Hamburg (1899–1976) was a pioneer in California agriculture who transformed the San Joaquin Valley from a semi-arid desert to a fertile agricultural region. He spent many years as an agricultural advisor in Israel to the governments of David Ben Gurion and Golda Meir. This illustrated work, published in conjunction with the Symposium "Ethnicity in California Agriculture" and the Western Jewish History Exhibit from the Hamburg Archives, contains a tribute to Hamburg and his impact on agricultural development in California and Israel. It consists of several sections:

1. Author Acknowledgements (pp. 7–8).
2. Foreword by Moses Rischin, Director, Western Jewish History Center (pp. 11–12.
3. "Sam Hamburg, Eloquent Man of the Soil," by Alice Hamburg (Wife)—A biography of Hamburg covering his childhood in Russia, his years in Palestine, and his extraordinary career as a prominent farmer, advisor, and agricultural legend (pp. 13–53).
4. Inventory of Hamburg Archives—Guide to letters, correspondence, papers, scholarships, articles, photos and other materials in the Hamburg Archives at the Western Jewish History Center (pp. 55–67).
5. Memorial essay by Lionel Steinberg, President of David Freedman Company, Thermal, CA. (1989)—Steinberg, a personal friend of Hamburg, reminisces about Hamburg and the people that knew him (pp. 71–74).
6. Memorial speech (1976)—Eulogy presented by Fresno Cantor Michael Loring (pp. 75–76).
7. Quotations about Hamburg from American and Israeli leaders (pp. 77–79).
8. Excerpts from letters written by Hamburg to Golda Meir and David Ben Gurion, and letters to Hamburg from Teddy Kollek (Mayor of Jerusalem) and W. Zev Hassid, a friend and Professor at UC Berkeley (pp. 79–83).
8. Western Jewish History Center Acknowledgments (p. 85).
9. Printed program for the Symposium on "Ethnicity in California Agriculture" (p. 87).

262. Herscher, Uri D. (1981). *Jewish Agricultural Utopias in America, 1880–1910.* Detroit: Wayne State University Press. 197 pp. ISBN 0814316786.

During the mass migration of Jews from Eastern Europe in the 1880s, Jewish organizations established various programs to settle immigrants on individual farms and farming communities around the country. This book presents an overview of settlements in New York, Louisiana, South Dakota, Arkansas, Colorado, and New Jersey and their unique social, cultural, and economic conditions. Many of the colonies were out of existence by 1910; New Jersey colonies remained viable through the 1950s. Herscher gives reasons for the failure of the settlements and his ideas on why the New Jersey colonies succeeded for a longer period. He compares colonization efforts in Israel and America and contrasts the failure of settlements in Ottoman Palestine and the successes of the kibbutz movement. The appendix covers excerpts from *The Jewish Farmer* regarding the New Jersey Colonies and a memoir written by Sidney Bailey a founder of the Alliance, New Jersey settlement.

263. Joseph, Samuel (1935). *History of the Baron De Hirsch Fund: The Americanization of the Jewish Immigrant.* Philadelphia: Jewish Publication Society. 322 pp. ISBN. (Reprinted with a new foreword by Philip S. Cohen and Francesco Cordasco. Fairfield, NJ, Augustus M. Kelley, 1978.)

A history of the creation, policies, and activities of the Baron De Hirsch Fund and its role in helping immigrants adjust to life in America. The fund was instrumental in establishing the Woodbine Agricultural settlement, the Jewish Agricultural Society, the Agricultural School, the Industrial Removal Office, a trade school and various other immigrant aid activities. Explores the reasons for the mass migrations, the effects of the settlements on Jewish life, and the Fund's impact on Jewish involvement in American life in the early 20th century.

264. Kann, Kenneth (1993). *Comrades and Chicken Ranchers: The Story of a California Jewish Community.* Ithaca, N.Y.: Cornell University Press. 313 pp. ISBN 0801428076.

In the early 1900s, a small group immigrants from Eastern Europe settled in Petaluma, California and established a poultry farming community raising eggs and chickens to serve the growing population in the area. This ethnic rural community was designed with a socialist orientation and a Jewish life style. On the 50th anniversary of the Petaluma Jewish Community Center (1975), Kann began collecting oral interviews with first, second, and third generation settlers to gain insights into their lives and what it was like to be part of this unique farming community. He presents excerpts and recollections from some two hundred interviews covering settler attitudes about religion, politics, culture, community tensions, Jewish identity, Hebrew and Yiddish, Zionism, relations with Gentiles, family farming, and other aspects of their life in Petaluma. This book gives readers a fascinating glimpse of life in a Jewish farming community and the changing ideological and cultural differences among the three generations.

265. Kann, Kenneth (1981). *Joe Rapoport: The Life of a Jewish Radical.* Philadelphia: Temple University Press. 315 pp. ISBN 0877222088.

Joe Rapoport was an immigrant from Russia and a dedicated socialist who became a knit goods worker, trade unionist, and political activist. This autobiography based upon oral interviews over several years, traces Rapoport's life, his views and his many activities in support of civil and human rights. Rapoport and his wife Sheba moved to Petaluma, California in 1949 to join the community of chicken farmers. In a riveting narrative, he describes the community, his orientation to the new life, and the years of social activism. Kann discusses the ideological conflicts among the farmers, the toll that red baiting took on the community during the years of Senator McCarthy's anti-communism crusade, and the sense of solidarity and cohesion when the community came together to support causes they believed in. With the corporate takeover of egg and poultry production in the 1950s and 1960s, many farmers were forced out of business. Rapoport remained in chicken farming until 1973 when he sold his farm. He remained in the area and continued to be active in Democratic party politics.

266. Klein, Moses (1889). *Migdal Zophim (The Watch Tower): The Jewish Problem and Agriculture as Its Solution.* Philadelphia: Published by the author. 125 pp.

In this two part work, written in English and Hebrew, Klein advocates encouraging emigrants from Russia to settle in agricultural colonies in their new countries. He begins by describing the poverty and oppression facing Jews in Russia in the 1880's. He was concerned that Jewish youth in Russia at the time were not getting proper schooling. Jews were leaving Russia in large numbers and emigrating to the United States and Palestine. Drawing upon articles from *The Jewish Exponent* (December 16, 1887), he describes life in Palestine and the efforts by Sephardim, Ashkenazim and Pharisees to build a country and produce an abundant agriculture. Klein then focuses on Jewish agricultural colonies in America and particularly the three New Jersey colonies of Alliance, Rosenhayn and Carmel from articles in *The Philadelphia Sunday Mercury* (October 20 and 27 1889). He provides details and statistics and some photographs about the settlers and their skills, history, buildings, schools, and synagogues.

267. Lee, Samuel James (1970). *Moses of the New World: The Work of Baron De Hirsch.* New York: Thomas Yoseloff. 313 pp. ISBN 0498073785.

This extensive biography explores the life and work of Baron Maurice De Hirsch (1831–1896). Covers his childhood in Germany, his family, his travels, business interests, and philanthropic ventures. Emphasizes Hirsch's concern about the oppression of Jews in Eastern Europe and his desire to help Jews living in those societies to establish better lives in democratic countries. Through philanthropic work, diplomacy, and influential organizations, Hirsch's efforts enabled Jews to settle in England, Argentina, and North America. Chapter 18, "North America," is devoted to activities regarding the placement of Russian immigrants in agricultural settlements throughout the United States and Canada. The Baron De Hirsch Fund and the Jewish Agricultural Society, supported settlements in New Jersey, New York, Louisiana, Colorado, Michigan, Connecticut and other locations. Lee explores the settlement efforts and the successes and failures of the program.

268. LeMaster, Carolyn Gray (1994). *A Corner of the Tapestry: A History of the Jewish Experience in Arkansas.* Fayetteville, AR: University of Arkansas Press. 645 pp. ISBN 1557283044.

A history of the Jewish experience in Arkansas from 1823 through the 1990s. Data taken from American Jewish Archives, periodicals, newspapers, interviews, and letters. Covers the pioneer settlers, the immigrants from Germany and Eastern Europe and the economic and civic progress achieved by the Arkansas Jews. Jewish involvement in Arkansas agriculture was primarily in cotton raising, handling, and trading. Includes account of a failed attempt by Russian immigrants from the Am Olam movement to establish an agricultural colony near Newport, Arkansas. The colony was abandoned after one year due to unexpected problems of climate, economic deprivation, illness, snakes, mosquitoes, and malaria.

269. Levine, Rhonda F. (2001). *Class, Networks, and Identity: Replanting Jewish Lives from Nazi Germany to Rural New York.* Lanham, MD: Rowman & Littlefield. 206 pp. ISBN 0742509923.

A sociological study of refugees from Nazi Germany who became successful cattle dealers in south central New York. Levine explores their lives in rural Germany, the lure of rural New York for their new homes and how they created a successful cattle industry in the region. Using data gathered from personal interviews and primary and secondary sources, Levine gives details of the lives of the cattle dealers and their home, work, and religious lives, the role of women, and relationships with Christian neighbors and the larger Jewish community. Changes in the cattle industry and the attraction of greater opportunities elsewhere led to the decline of the community in the late 1960s as most of the second generation left the region to settle in urban areas. Levine concludes by comparing the experiences of the German Jewish immigrants with the adaptation and mobility of immigrants in general.

270. Makover, Abraham B. (1917). *Mordecai M. Noah, His Life and Work from the Jewish Viewpoint.* New York: Bloch Publishing Co. 96 pp.

Mordechai Manuel Noah was a lawyer, diplomat, philanthropist, and statesman. He was born in Philadelphia in July 1785 and died in New York City in March, 1851. Makover gives readers a view of Noah's life and activities and the wide range of interests and causes on behalf of Jews around the world to live free from oppression and fear. Much of the book examines Noah's plan to establish "Ararat" a refuge for Jews on Grand Island near Buffalo, New York. Noah's plan was widely criticized by Jewish leaders. A cornerstone was laid in Ararat but Noah was unable to find settlers willing to live there and the plan failed. Nevertheless, Makover depicts Noah as a visionary and dedicated worker for Jewish liberation.

271. Meckler, Brenda Weisberg (1988). *Papa Was a Farmer.* Chapel Hill, NC: Algonquin Books in Association with Taylor Publishing. 318 pp. ISBN 0912697954.

A biographical memoir of Meckler's life growing up on a farm in Ohio during the years 1906–1917. Her parents migrated to the United States, settled briefly in Boston and three years later bought a farm on Dunham's Hill, Ohio, 25 miles from Cincinnati. Meckler recounts her experiences on the farm, her parents' perseverance and their desire to maintain their Yiddish language, Jewish life, and customs in the isolated community. She tells about her friends, education, farm work, and the joys of country life. The farm was sold in 1917 and the family moved to Phoenix.

272. Monaco, C.S. *Moses Levy of Florida: Jewish Utopian and Antebellum Reformer.* Baton Rouge, LA. Louisiana State University Press, 2005. ISBN 0807130958. 251 pp.

Biography of Moses Elias Levy, a Moroccan born intellectual Jew who become a landowner, shipping magnate, social activist, and abolitionist. Levy had land holdings over 100,000 acres where he raised sugar, tropical fruits, roots, and seeds and was described by some as an agricultural innovator. He advocated the establishment a colony of refuge for European Jews on his Plantation Colony in Florida which did not receive wide support from Jewish leaders of the time. Levy takes readers on a journey through the life of a remarkable 19th century reformer who worked for justice for Jews, supported interfaith cooperation, and participated in Florida's development. (See also: Monaco, Chris [2002]. A Sugar Utopia on the Florida Frontier: Moses Elias Levy's Pilgrimage Plantation. *Southern Jewish History,* 5[1–5]: 103–140.)

273. Navarra, Tova (1996). *Howell and Farmingdale: A Social and Cultural History.* Arcadia Publishing: Dover, NH. 128 pp. ISBN 0752402838.

A photographic history of the people, places, and community events in the townships of Howell and Farmingdale, New Jersey from 1965 to 1995. The visual images and accompanying text show the community and its diversity, education, and social life during the 30 years Navarra lived there. Chapter 5, "Industry, farming and business" (pp. 99–128), includes scenes of Jewish chicken farmers and their families.

274. Norman, Theodore (1985). *An Outstretched Arm: A History of the Jewish Colonization Association.* London: Routledge & Kegan Paul. 342 pp. ISBN 0710202539.

The Jewish Colonization Association was established in 1891 with funds from the estate of Baron Maurice de Hirsch to settle Russian Jewish immigrants in agricultural settlements in Argentina, Brazil, Canada, Turkey Cyprus, and Israel. Norman presents the many faceted personality of Hirsch, his philanthropic efforts on behalf of Jewish settlement and the Association's impact in the respective countries. Includes only brief coverage of American settlement with particular reference to the New Jersey colonies (pp. 108–111).

275. Nudelman, Eugene R. (1969). *The Family of Joseph Nudelman.* Published by Eugene R. Nudelman. 30 pp.

Joseph Nudelman came to America in 1881 with a desire to live his life on the land in an agricultural settlement. This privately published book by his great grandson is an inspiring story of Nudelman's life as an agricultural pioneer in the late 19th and early 20th century. In 1883, Nudelman and his family settled with other families in the Painted Woods Colony in the Dakota Territory. When Painted Woods failed in 1885 they moved to Portland, Oregon. A year later he bought land near Porterville, Folsom, San Reymon, and Heywood, California and some years later moved to Nevada to try his hand at ranching. At age 58, Nudelman sold his ranch in Nevada and moved back with his family to Portland, Oregon where he opened a kosher meat market and other businesses. Nudelman died in 1935. (For more information on the Dakota years told by Nudelman's great-great-grandson see: Fields, Harvey J. [2002] Joseph Nudelman: the Dakota Years, 1844–1934. *Western States Jewish History,* 35[1]: 9–19.)

276. Ornstein-Galicia, Jacob (1992). *Jewish Farmer in America: The Unknown Chronicle.* Lewiston, NY: E. Mellen Press. 252 pp. ISBN 0773492291.

Ornstein-Galicia presents a chronicle of Jewish farming in America from 1908 through 1960. His parents arrived in America in 1908 and settled on a farm in Northeastern Ohio where he grew up. In part I, the author gives a historical sketch of Jewish farmers in the 19th and 20th centuries. Part 2 offers "vignettes" of the typical life of an American Jewish farm family. The vignettes depict the family, the farms, the community, relationships with relatives and neighbors, and the joys and frustrations of living and working on a family farm in Ohio.

277. Patt, Ruth Marcus (1982). *The Jewish Scene in the Central Jersey Rural Areas: Supplement No. 2. Supplement to: The Jewish Scene in New Jersey's Raritan Valley, 1698–1948.* New Brunswick, NJ: Jewish Historical Society of Raritan Valley. 36 pp.

A history of the early families and new settlers, congregations, educational institutions, businesses, social services, Zionist activity, and everyday lives of the Jews of Central New Jersey. This supplement devoted to the Jewish farmers, gives a brief history of the agricultural colonies in central new Jersey; biographies of farmers and their families, professors, scientists and agricultural extension agents at Rutgers University, and an excerpt from the diary of Abraham Dobin, author of *Fertile Fields* (South Brunswick, A.S. Barnes, 1975) regarding his work as an agricultural extension agent.

278. Patt, Ruth Marcus (1994). *Uncommon Lives: Eighteen Extraordinary Jews from New Jersey.* New York: Vantage Press. 211 pp. ISBN 053310970.

Contains detailed biographies of famous Jewish personalities who made significant contributions to medicine, education, law, politics, art, and business in New Jersey. Includes biography of Jacob Goodale Lipman (p 51–63), a Rutgers graduate who became an agricultural chemist and professor at Rutgers College of agriculture (1910), director of the Rutgers Agricultural Experiment Station (1911) and Dean of the College of Agriculture (1915–1939), and Selman Waksman (p.77–86), an agricultural scientist, first Director of the Institute of Microbiology at Rutgers and recipient of the Nobel Prize for discovery of Streptomycin.

279. Richman, Irwin (1998). *Borscht Belt Bungalows: Memories of Catskill Summers.* Philadelphia: Temple University Press. 251 pp. ISBN 1566395852.

Richman spent summer vacations with his family at bungalow colonies in the Catskill region of New York State also known as the "Borscht Belt." He takes readers through a history of the region and its growth from a quiet rural area of New York State to a resort serving nearly a million vacationers every summer. He describes the region, the land, the towns and villages, the bustling daily life of the summer residents, and the nature of Jewish life in the region. Richman remembers the Jewish farmers and their impact on the region as developers of bungalow colonies and suppliers of food to the resorts. He offers his observations about cultural and social life in the bungalow colonies and the famous resort hotels and the changes which led to the decline of the region as a vacation area.

280. Rochlin, Harriet, and Fred Rochlin (1984). *Pioneer Jews: A New Life in the Far West.* Boston: Houghton Mifflin. 262 pp. ISBN 0395318327.

A detailed history of Jewish pioneers who settled in western United States as early as the 16th Century and made an impact on their communities and their professions. Covers Jewish pioneers in the gold rush, merchants and industrialists, elected and appointed officials, and famous Jewish families. Includes farmers, cattle raisers and land speculators in Arizona, California, Colorado, Utah, and Wyoming.

281. Rosengarten, Theodore, and Dale Rosengarten (2002). *A Portion of the People: Three Hundred Years of Southern Jewish Life.* Columbia, SC: University of South Carolina Press in association with McKissick Museum. 286 pp. ISBN 1570034451.

This catalog containing essays and photographs, was issued in conjunction with an exhibition on three hundred years of Jewish life in South Carolina. It tells the story of the South Carolina Jews and their lives, their work, and their religious practices. An essay by Jenna Weissman Joselit (pp. 22–30) gives a background on the Russian Jews who migrated to the state from 1907 to 1920 and how they became peddlers, shopkeepers and farmers. In 1907, twenty five Russian Jewish families established an agricultural cooperative on 2200 acres of land outside Aiken. The settlement, called Happyville, had a life of only two years due to unusually cold weather, lack of agricultural know how, and financial setbacks. Settlers left for New York, New Jersey and other areas of the South. Mentions other South Carolina Jews who were cotton planters and landowners. The photo essay by Bill Aron includes a photo of the largest squab farm in the world in Sumter, SC, operated by Wendell M. Levy and his twin sister Patty Levi Barnett.

282. Rosskam, Edwin (1972). *Roosevelt, New Jersey: Big Dreams in a Small Town & What Time Did to Them.* New York: Grossman. 148 pp. ISBN 0670607517.

Roosevelt, New Jersey originally called "Jersey Homesteads," was established in 1936 by Jewish immigrants from Eastern Europe who established a cooperative garment factory and cooperative farms as their main sources of employment. They wanted the town to serve as a model for preserving Jewish culture and the Yiddish language in a socialist setting. By the end of WWII the cooperative ventures had ended and the name of the town was changed to 'Roosevelt' in memory of President Franklin D. Roosevelt. Over time it developed into a typical rural town with a predominantly Jewish population that included writers, artists, musicians, and tradesman. In a flowing narrative Rosskam presents a memoir of Roosevelt and the people and places that made the town unique.

283. Rubin, Harris (1977). *The Autobiography of Harris Rubin.* Philadelphia: Bricklin Press. 158 pp.

Harris Rubin was born in Lithuania in 1847 and came to the United States in 1882, his family joined him in 1885. In this memoir, written during a five year period from 1915–1920, Rubin describes what it was like growing up in Lithuania, and his desire to make a new life in America as a farmer. In America he held several farm jobs, and eventually settled in the Jewish agricultural settlement in Carmel, New Jersey and opened a butter and egg business in Philadelphia. Rubin gives his impressions of life in America and his experiences as a farmer and businessman. This memoir, translated from the Yiddish, offers readers a glimpse of the thoughts and daily life of an immigrant and his family and their adaptation to the new country.

284. Sabsovich, Katharine (1922). *Adventures in Idealism: A Personal Record of the Life of Professor Sabsovich.* New York: Privately printed for the author by Stratford Press. 216 pp. (Reprinted Arno Press, 1975. ISBN 0405067453.)

Biography of Hirsch Leib Sabsovich, an agricultural chemist, administrator of the Baron D. Hirsch Fund, and founder of the Jewish agricultural settlement at Woodbine, New Jersey. This biography, written by his wife after his death, begins with a description of life in Russia in the early 20th Century and their immigration to the United States to make a new life. Sabsovich held a brief appointment as Professor at Colorado State University and then was offered the position of General Manager of the Baron D. Hirsch Fund and became instrumental in establishing the Jewish agricultural colony at Woodbine, New Jersey. The author recounts life with her husband and characteristics of the settlers, the progress and development of the colony, and establishment of an agricultural school. She writes about the land and farming techniques at Woodbine, social relationships with others in the community, education programs, and the disappointments, hardships, and joys of those years. Part II "From Those Who Knew Him Best" contains seven personal tributes to Sabsovich from individuals associated with him at the Hirsch Fund, the agricultural school, and Woodbine.

285. Schloff, Linda Mack (1996). *And Prairie Dogs Weren't Kosher: Jewish Women in the Upper Midwest since 1855.* St. Paul, MN: Minnesota Historical Society Press. 253 pp. ISBN 0873513371.

Explores the lives of four generations of Jewish women who settled in Minnesota, North Dakota, South Dakota, parts of Iowa, and Northern Wisconsin. Many of the women lived with their families on homestead land from 1880 to 1920; others had parents or spouses involved in related businesses. Schloff shows how these courageous women adapted to life on the prairies and small towns of the upper Midwest and the roles they created for themselves as they eventually moved into the middle class. The data is based upon oral accounts, family histories, organizational records, and photographs. The women offer anecdotes and reminiscences of life in the old country, their lives in America, domestic life, farm work, and their involvement in synagogues and organizations. This work is a tribute to generations of women who shaped their families and communities and made important contributions to Jewish life in the upper Midwest.

286. Schoener, Allon (1983). *American Jewish Album: 1654 to the Present.* New York: Rizzoli. 342 pp. ISBN 084780500X.

Schoener takes a journey through the American Jewish experience from 1654 to the present with 500 illustrations and text from journal entries, oral interviews, and personal letters covering Jews in every aspect of American life. He captures the Jewish agricultural presence with illustrations and some text from settlements in Bismarck, North Dakota, Woodbine, New Jersey, Cotopaxi, Colorado and fur traders in Alaska.

287. Tenenbaum, Shelly (1993). *A Credit to Their Community: Jewish Loan Societies in the United States, 1880–1945.* Detroit, MI: Wayne State University Press. 204 pp. ISBN 0814322875.

An analysis of Jewish free loan societies and their impact on the communities they served during the periods of Jewish immigration to America. Explains the cultural settings and values of the societies and the immigrants' involvements with them. Includes brief coverage of credit unions serving Jewish farmers and the role of Jewish Agricultural and Industrial Aid Society in promoting credit unions.

288. Toll, William (1982). *The Making of an Ethnic Middle Class: Portland Jewry over Four Generations.* Albany, NY: State University of New York Press. 254 pp. ISBN 0873956095.

A study of the continuity and social change among the Jewish community of Portland, Oregon from 1830 through the present. Although the Jews of Portland were largely a community of families in mercantile and craft professions, some were supporters of communal organizations and were attracted to agriculture and rural life. Toll discusses the ideas and motivations of settlers that left Portland to pursue a life in farming in colonies in Oregon and the Dakota Territory or as independent farmers. Other families from rural Minnesota, Missouri, and Kansas moved to Portland in later years and became active in Portland civic affairs and the Jewish community.

289. Trupin, Sophie (1984). *Dakota Diaspora: Memoirs of a Jewish Homesteader.* Berkeley, CA: Alternative Press. 160 pp. ISBN 0960634649. (Reprinted: Lincoln, NE, University of Nebraska Press, 1988. ISBN 080329414X.)

Trupin recounts her childhood growing up on a farm in the Dakota wilderness in the early 1900s. She begins with an account of her parents' life in Russia and their eventual immigration to America and the journey to the North Dakota homestead. Trupin tells about the isolation and the poor land and climate which made farming difficult, her school and friends, the community, and relationships with other Jewish farmers and non-Jewish neighbors. Despite the lack of a rabbi, synagogue, kosher butchers, and religious school, Trupin's family found ways to retain their Jewish customs and observe holidays and rituals. Her parents left farming in 1953 and moved to Israel where they spent their last years.

290. Waksman, Selman A. (1966). *Jacob G. Lipman: Agricultural Scientist and Humanitarian.* New Brunswick, NJ: Rutgers University Press. 152 pp. (Reprinted 1981.)

Jacob Goodale Lipman (1874–1939), the son of Russian Jewish immigrants grew up in the Baron D. Hirsch agricultural colony in Woodbine, New Jersey. Lipman studied soil science and agriculture at Rutgers and Cornell universities. He was appointed to the faculty at Rutgers College of Agriculture in 1901 where he established the Department of Soil Chemistry and Bacteriology. He was a founder of the prestigious journal *Soil Science* and in 1915 was appointed Dean of Agriculture and Director of the New Jersey Agricultural Experiment Station. Waksman draws upon his 30 year association with Lipman, records, personal diaries, and recollections from those who knew Lipman, to document the life of a renowned agricultural scientist who made important advances in agriculture worldwide. Includes excerpts from Lipman's diaries.

291. Wolin, Penny Diane (2000). *The Jews of Wyoming: Fringe of the Diaspora.* Cheyenne, WY: Crazy Woman Creek Press. 198 pp. ISBN 0967635705.

A collection of over 100 photographs from a traveling exhibition documenting the lives of Jews in Wyoming from 1860–2000. The photographs depict Jews in their homes, their work, their synagogues and religious observances with accompanying text describing Jewish life in Wyoming in their own words. Includes photos of farmers, ranchers, and cowboys.

Book Sections

292. Adler, Joseph G. (1984). Moses Elias Levy and Attempts to Colonize Florida. In: *Jews of the South: Selected Essays from the Southern Jewish Historical Society.* Samuel Proctor, Louis Schmier and Malcolm Stern, editors. Macon, GA: Mercer University Press. Pp. 17–29.

Moses Elias Levy was the son of a Moroccan born government minister who believed that Jewish immigrants in the United States would gain acceptance from their gentile neighbors by choosing a life on the land in religious agricultural settlements where they can live as Jews and avoid assimilation. Levy traveled widely through Europe and the United States trying to gain support for his settlement idea. He acquired land in Florida and established several plantations but was unable to attract a large number of settlers. Levy abandoned the idea by the 1830s and sold the land to recoup his investments. One of his plantations was destroyed in 1835 during the Second Seminole War. Adler presents an analysis of Levy's crusade and some views on why it failed.

293. Ard, Patricia M., and Michael A. Rockland (2002). Jews on the Farm. In: *The Jews of New Jersey: A Pictorial History.* New Brunswick, NJ: Rutgers University Press. ISBN 0813530121. Pp. 59–75.

A photographic essay depicting representative Jewish communities throughout New Jersey and the daily lives of their inhabitants. Chapter 5, "Jews on the Farm," presents a history of the former Jewish agricultural settlements and individual farms in Alliance, Carmel, Norman, Farmingdale, Roosevelt, and other communities. Captioned photographs and accompanying text depicts life on the farm, community activities, personal relationships and memorable events.

294. Bartelt, Pearl (1997). American Jewish Agricultural Colonies. In: *America's Communal Utopias.* Donald E. Pitzer, editor. Chapel Hill: University of North Carolina Press. ISBN 080782299X. Pp. 352–374. (Reprint of a paper presented at National Historic Communal Societies Association Annual Meeting. 1989.)

A history of the Jewish agricultural colony movement in the 19th and early 20th centuries. Bartelt focuses on the early Russian immigrant communities and the colonies formed at the turn of the century in New York, New Jersey, North Dakota, South Dakota, Florida, Colorado, Kansas, Louisiana, Oregon, Utah and Michigan. She explains the forces that motivated the formation of the colonies and the lives and experiences of the immigrants who settled in them. She points out that although the colonies were short lived they helped shaped the cultural geography of their respective regions.

295. Bitensky, Samson (2002). Moses Elias Levy and a Jewish Zion in Florida. In: *Blessings of Freedom: Chapters in American Jewish History.* Michael Feldberg, editor. Hoboken, NJ: KTAV Publishing House in association with The American Jewish Historical Society. ISBN 0881257559. Pp. 204–205.

The story of Moses Elias Levy and his plan to provide a home in Florida for the persecuted Jews of Europe. In 1821, Levy, a munitions contractor and shipping magnate, purchased 53,000 acres of land near Micopany in northeastern Florida to establish a colony called which he called "Pilgrimage Plantation." From 1820–1824 he traveled widely seeking support from Jews and Christians in the United States and England by attempting to show that persecuted Jews need a place to live in peace. Although Levy impressed his audiences with his erudition and knowledge of Judaism, his pleas for support for the colony were largely ignored. In 1835, the plantation was destroyed by fire during the Second Seminole War and the plan was abandoned. During his visit to England, Levy persuaded the British government with the support of upper class Christians to restore full political equality to England's Jews. Bitensky observes that Levy's legacy of a home for Jews in Florida has been realized by the fact that Florida comprises the third largest concentration of Jews in the United States.

296. Bogen, Boris D. (1917). The Back to the Soil Movement. In: *Jewish Philanthropy: An Exposition of Principles and Methods of Jewish Social Service in the United States.* Boris David Bogen, editor. New York: Macmillan. ISBN 0875850863. Pp. 125–137. (Reprinted Montclair, NJ Paterson Smith, 1969. ISBN 0875850863.)

A brief survey of philanthropic efforts from 1837 to 1916 designed to entice new immigrants to pursue a life on the farm in agricultural settlements. Bogen outlines the work of American Hebrew Agricultural Association, Alliance Israelite Universelle, Immigrant Aid Society of New York, Agricultural and Industrial Aid Society, and the Baron De Hirsch Fund. These philanthropic groups provided loans and mortgages, established agricultural schools, farmers federations, and newspapers. Local groups like the Jewish Agricultural and Colonial Organization (Philadelphia), Utah Colonization Fund (Salt Lake City) and the Jewish Farmers Association of St Louis supported settlement efforts in their regions. Bogen gives statistics on the demography, land values, and economic considerations that each colony had to consider. He discusses the Woodbine (NJ) settlement and the features (i.e. factory, machine shop, self government, farm school) that enabled it to survive. The section concludes with Morris Loeb's observations regarding the elements necessary for an agricultural colony to succeed and prosper. (See also pages 27–31 for more on the Baron De Hirsch Fund.)

297. Brown, Phil (1998). How the Jewish Catskills Started. In: *Catskill Culture: A Mountain Rat's Memories of the Great Jewish Resort Area.* Philadelphia: Temple University Press. ISBN 1566396425. Pp. 23–44.

Catskill Culture is a story of the Catskill resort area (also known as "The Borscht Belt) and its memorable years, its unique qualities, the factors that led to its decline and changes in the ethnic character beginning in the 1970s. The Catskills, an area of nearly 4 million acres, comprises Greene, Delaware, Sullivan, Ulster, Schoharie, Otsego and part of Albany counties. Many of the resorts had their origin as farms owned by Jews who began taking in boarders to supplement their farm income. Brown discusses the agricultural legacy of the region which began at the end of the 19th century, the role of the Jewish Agricultural Society in helping to establish the farms and the cooperative ventures that helped keep the farmers in business. By 1910, there were 741 farm/boarding houses in the region, the largest concentration of Jewish farmers in the country. Over the years these farms developed into famous country clubs and resorts that served as getaways for those seeking a respite from urban life. Brown explains how Jews got started in farming, why they chose the Catskills, and how their presence influenced other Jews to settle there and transform the area into a famous vacation spot.

298. Davidson, Gabriel (1941–1943. Colonies, Agricultural: III. United States. In: *The Universal Jewish Encyclopedia: An Authoritative and Popular Presentation of Jews and Judaism since the Earliest Times.* Isaac Landman and Louis Rittenberg, editors. New York: Universal Jewish Encyclopedia. Pp. 294–297. (Reissued New York, Ktav, 1969.)

Brief history of the Jewish agricultural colonies established in the late 19th and early 20th centuries. The 19th century colonies were established in New York, Louisiana, North Dakota, Oregon, Kansas, New Jersey, Virginia, Michigan, and California. Attempts to start colonies in the early 20th century occurred in Maryland, Alabama, Texas, Georgia, Tennessee, Michigan, and North Carolina. Many of the settlements were short lived; they failed after a few years due to a variety of economic and social difficulties. Some Jews who set up farms individually were thriving and the Jewish Agricultural Society estimated that there were about 100,000 Jews engaged in agriculture in 1940. Some who grew up on the farms, became teachers or staff of agricultural schools, experiment stations, and state or federal agencies. (See also: Reizenstein, Milton. Agricultural colonies in the United States. *The Jewish Encyclopedia*, New York, Funk and Wagnalls, 1906, pp. 256–262; and *Encyclopedia Judaica* Jerusalem, Keter, 1971, Volume 2, pp. 411–413.)

299. Fein, Shirley P. (1996). Down on the Farm. In: *Nice Jewish Girls: Growing Up in America.* New York: Plume. ISBN 0452273978. Pp. 143–146.

Reminiscences of growing up as a girl on a farm in Northeastern Connecticut in the 1920s. Fein's grandfather, a refugee from Russia, lived with the family. She gives brief accounts of the enjoyable experiences on the farm and the disappointments she had to bear from a grandfather who had negative attitudes regarding girls working, driving, or receiving an education.

300. Feldstein, Stanley (1978). Return to the Soil and the Rural Experience. In: *The Land That I Show You: Three Centuries of Jewish Life in*

America. Stanley Feldstein, editor. New York: Anchor Press/Doubleday. ISBN 0385024452. Pp. 49–59, 123–146.

In the first half of the nineteenth century, there was a major effort to settle new immigrants in agricultural co-operatives around the country. Feldstein describes these efforts as a combination of American romanticism and Jewish messianism. He summarizes proposals for settlement in the Mississippi Valley, Florida; Shalom Colony and Ararat Colony (Grand Island) New York; Sicily Island, Louisiana and colonies in The Dakotas, Oregon, Colorado, New Jersey, Wisconsin, Utah and Arkansas. Some of the efforts failed and the plans were dropped. The colonies survived about 3–5 years because they were located in areas poorly suited for farming with inclement weather, floods, and crop failures. Most settlers were untrained or lacked farming experience and were overwhelmed by debts and conflicting ideologies. The New Jersey colonies eventually abandoned cooperative ventures and emphasized individual ownership. They survived through the early twentieth century possibly because of access to the New York markets and their mixed agricultural/industrial settling which provided steady incomes. Feldstein concludes that the settlements were destined to fail because the urban and industrial revolutions afforded immigrants the option of abandoning rural life for opportunities in the urban sector.

301. Fishberg, Maurice (1911). Jews as Agriculturists. In: *The Jews: A Study of Race and Environment.* New York: Arno Press. ISBN 0405067100. Pp. 403–406, 499.

Brief history of Jews in agriculture through the ages. Includes the agricultural colonies established in the United States. Fishberg indicates that there are 3040 Jewish farmers in the United States with the largest number in New Jersey, Connecticut, New York, and North Dakota.

302. Frank, H. (1955). Jewish Colonization in the Americas: I. Jewish Farming in the United States. In: *The Jewish People, Past and Present. Volume 2.* New York: Jewish Encyclopedic Handbooks. Pp. 69–77.

A look at Jewish farming in the United States during the late 19th and early 20th centuries. Covers the early colonization supported by the Baron de Hirsch Fund and the Jewish Agricultural Society; full and part time farming by individual farmers; post-war settlement of veterans on farms and geographic distribution of the Jewish farm population. Gives details regarding kinds of farming, characteristics of farmers, cooperative ventures, economic situation, and farm life.

303. Friedman, Lee A. (1942). Ararat—A City of Refuge for the Jews. In: *Jewish Pioneers and Patriots.* Philadelphia: Jewish Publication Society of America. Pp. 107–115.

A biography of Mordechai Manuel Noah, journalist, playwright, author, and politician. Describes Noah's crusade to establish a colony called "Ararat" at Grand Island, NY on the Niagara River to serve as a refuge for Jews from all parts of the world. This chapter explains Noah's efforts to establish Ararat and the motivations of his supporters and detractors. Although a cornerstone was laid, the colony failed to attract settlers and the plan failed. Grand Island was sold and Noah moved on to other activities.

304. Goldberg, Robert A. (1991). Zion in Utah: The Clarion Colony and Jewish Agrarianism. In: *Jews of the American West.* Moses Rischin and John Livingston, editors. Detroit: Wayne State University Press. ISBN 0814321704. Pp. 67–94.

A history of the Jewish agricultural Colony at Clarion, Utah. The Clarion Colony was founded in 1911 at a site 135 miles southwest of Salt Lake City with assistance from the Jewish Agricultural and Colonial Association. The settlement was based on an idealistic back to the land movement operating as a cooperative with socialist ideals. Poor land, inexperience, lack of economic stability, meager resources, and low morale led to the demise of the colony by 1915. Goldberg explores the land, motivations of settlers and reasons for the failure of Clarion and settlements established in other states. He points out that the agricultural societies that helped to finance and establish the colonies were too hasty in choosing sites and underestimating costs and not doing enough to train the new immigrant settlers for life on the land.

305. Goodstein, Phil (1992). Pioneering Agricultural Colonies. In: *Exploring Jewish Colorado.* Denver, CO: Rocky Mountain Jewish Historical Society: Center for Judaic Studies, University of Denver. Pp. 80–85.

Description of the Cotopaxi Colony established in 1882 and the Atwood Colony established in 1893. Both colonies were faced with economic problems, internal conflicts, inadequate housing and supplies and lack of farming experience. Cotopaxi lasted only two years and was abandoned in 1884. Atwood, established during the economic panic of 1893, existed for six years until 1889. Goodstein describes the founding of these colonies, the motivation of the settlers, the agencies that supported them, and reasons for their failure.

306. Grinstein, Hyman B. (1945). Agricultural Projects. In: *The Rise of the Jewish Community of New York, 1654–1860.* Philadelphia: Jewish Publication Society of America. ISBN 0879913517. Pp. 115–130, 453–460.

Analysis of the various projects aimed at settling Jewish immigrants in agricultural colonies. Covers Mordecai Noah and the Ararat project (1818), Society of Zeire Hazon (the Tender Sheep) in 1837, the attempts by the Lebanon Lodge of the International Order of B'nai Brith to establish a Hebrew Agricultural Society (1855), and the Sholem Colony in New York (1936). These efforts all failed to persuade immigrants to accept a communal life on the land. The chapter explores the philosophy behind the efforts and reasons for the lack of success. (See also pp. 453–460 for further details about Mordecai Manual Noah and the Ararat plan.)

307. Herscher, Uri D. (1975). Herman Rosenthal — The Jewish Farmer. In: *Michael: On the History of the Jews in the Diaspora.* Lloyd P. Gartner, editor. Tel Aviv: Diaspora Research Institute. Pp. Volume 3: 59–87.

In 1880, Herman Rosenthal (1843–1917) founded *Am Olam* in Ukraine with the goal of settling Jews on land in Palestine or the United States. He came to the United States in 1881 to carry out his plan and was successful working with American philanthropists to grant loans for establishment of agricultural colonies in various states. The New Jersey colonies — Alliance, Woodbine, Carmel, Rosenhayn and Vineland — succeeded because they combined agriculture and industry which provided a steady income from both ventures. In 1891, Rosenthal began publishing a Yiddish periodical called *Der Yudisher Farmer: Monatliche Tsaytshrift fir Landvirtshaftliche Kolonizatsyan* which lasted one year. This article presents seven excerpts, in Yiddish and English, translated by Herscher. The excerpts focus on the Woodbine and Alliance colonies with details about land, housing, and life in the settlements. The English translations appear on pages 59–74. The original Yiddish articles follow on pages 74–87.

308. Isaacs, Myer S. (1906). Jewish Farmers. In: *The Old Guard and Other Addresses.* New York: Knickerbocker Press. Pp. 176–183.

In this speech presented at the graduation of the Baron de Hirsch Agricultural School in Woodbine, New Jersey in March 1902, Isaacs explains the purpose of the School and its role in helping Russian and Roumanian immigrants adapt to agriculture. He responds to those who question whether Jews are suited for agriculture by referring to the many biblical accounts of Jews involved in agriculture and the present day movement of Jewish agricultural settlement.

309. Karp, Abraham J. (2002). Noah's Second Landing at Ararat. In: *Blessings of Freedom: Chapters in American Jewish History.* Michael Feldberg, editor. Hoboken, NJ: KTAV Publishing House in association with the American Jewish Historical Society. ISBN 0881257559. Pp. 206–207.

The story of Mordechai Manuel Noah and his plan to establish Ararat a city of refuge for Jews at Grand Island in the Niagara River between Buffalo and

Canada. Noah was a surveyor, a judge, and newspaper editor who issued proclamations to the Jews and Christians asking for support of his idea. He envisioned a city that would operate under the auspices of the United States and ruled by a judge elected every four years (Noah would occupy the office for the first four years). The plan, widely ridiculed by Jewish leaders and the press, failed to attract settlers to Ararat. Noah sold his share of the island to a timber investor in 1833. He continued to be a spokesman for American Jewry and the establishment of a Jewish State in the Holy Land.

310. Lavender, Abraham D., and Clarence B. Steinberg (2002). Jewish Farmers of the Catskills. In: *In the Catskills: A Century of the Jewish Experience in "the Mountains."* Phil Brown, editor. New York: Columbia University Press. ISBN 0231123604. Pp. 22–38.

In this excerpt from their book *Jewish Farmers of the Catskills* (University of Florida Press, 1995) Lavender and Steinberg present a short history and chronology of the growth of the Catskill area into a prosperous Jewish resort and farming community. They emphasize the importance of the railroad and boarding houses as factors that helped the region to grow and attract visitors and enabled the farmers to remain in business. The farmers established credit unions and farmers associations and built synagogues and religious schools to help preserve their Jewish life styles. Includes brief profiles of typical farm families and their lives in the Catskill area and relations with non-Jewish neighbors.

311. Levy, Abraham R. (1905). Rural Settlements: Western States. In: *The Russian Jew in the United States: Studies of Social Conditions in New York, Philadelphia, and Chicago; with a Description of Rural Settlements.* Charles S. Bernheimer, editor. Philadelphia: John C. Winston Co. Pp. 392–403.

During the summer of 1903, Levy and William Kahn, from the Jewish Agricultural and Industrial Aid Society (JAIAS), visited Jewish farmers in Dakota, Wisconsin, Michigan, Indiana, and Illinois. The purpose of the trip was to observe the conditions and progress of the farms and report back to the Society. Levy describes the families he visited, how they operate their farms, economic difficulties, and the impact of JAIAS support. Levy concludes that agriculture offers these settlers a way to achieve complete emancipation and achievement of a successful economic position in America.

312. Libo, Kenneth, and Irving Howe (1984). People of the Soil. In: *We Lived There Too: In Their Own Words and Pictures — Pioneer Jews and the Westward Movement of America, 1630–1990.* Kenneth Libo and Irving

Howe, editors. New York: St. Martin's/Marek. ISBN 0312858663. Pp. 282–298.

Personal accounts and anecdotes of the lives of pioneer Jewish farmers and homesteaders in North Dakota, Colorado, Wyoming, California, and New York during the late 19th and early 20th centuries gleaned from the Yiddish and English press.

313. Lipman, Jacob G. (1905). Rural Settlements: Eastern States. In: *The Russian Jew in the United States: Studies of Social Conditions in New York, Philadelphia, and Chicago; with a Description of Rural Settlements.* Charles S. Bernheimer, editor. Philadelphia: John C. Winston. Pp. 376–391.

History and analysis of the Jewish agricultural colonies of Woodbine, Carmel, Alliance, and Rosenhayn in Southern New Jersey and the individual Jewish farmers in the New England states of Rhode Island, Massachusetts, Connecticut, Maine, Vermont, and New Hampshire. Lipman discusses the characteristics of the colonies, the attitudes of the settlers regarding agriculture, education, politics, and religious observance. Jewish farmers in New England differ from the New Jersey farmers in their land, farming, and marketing conditions. They operate individually with a spirit of self reliance and independence and make their own decisions regarding their welfare, social life and religious observances.

314. Lowenstein, Steven (1987). New Odessa. In: *The Jews of Oregon, 1850–1970.* Steven Lowenstein, editor. Portland: Jewish Historical Society of Oregon. ISBN 0961978600. Pp. 89–91.

Details of the founding of the New Odessa agricultural colony in 1883 by about 50 Russian immigrants from the *Am Olam* movement. Explains how the land was acquired, the costs of land and building, philosophy of members, and the daily lives and activities at New Odessa. Differences of opinions and disputes about religious and secular issues, burgeoning debts, and a feeling of isolation led to bankruptcy and foreclosure and the colony was abandoned in 1888. The settlers left and most returned to New York. Eight members established a communal laundry, others became doctors, lawyers, pharmacists, dentist, engineer, chemist and teacher.

315. Macleod, Rob R. (1950). Noah's Architecture. In: *Cinderella Island.* Grand Island, NY: Rob Roy MacLeod. Pp. 25–29. (Reissued: Grand Island, N.Y.: Grand Island Chamber of Commerce, 1969. Includes 1855 photograph of Grand Island and the Ararat cornerstone.)

Brief history of the attempts by Mordecai M. Noah to establish an agricultural colony called Ararat at Grand Island as a city of refuge for Jews fleeing oppression in other countries. He intended it to be an independent self sustaining colony with an agricultural base. Noah purchased 2555 acres and laid a cornerstone with an elaborate celebration announcing the opening of Ararat. His efforts met with intense opposition and criticism and the project failed.

316. Manners, Ande (1972). The Man with the Hoe could be Jewish, Too! In: *Poor Cousins.* New York: Coward, McCann & Geoghegan. Pp. 135–172.

This book is about the lives and experiences of Russian Jews who came to America in the late 1800s. Manners explains how they adapted to their new country, raised their families and contributed to the economic, social, and religious life in America. Manners discusses the farming colonies in various states; the New Jersey colonies of Vineland and Alliance; and individual farmers in the Catskills who were instrumental in establishing the famous vacation resorts. Manners acknowledges the role of the Baron De Hirsch Fund and Jewish Agricultural Society as major players in the settlement efforts. He mentions some individuals from the colonies who had active careers in agricultural education and research.

317. Marcus, Jacob R. (1970). Agriculture, Ranching, Crafts and Professions. In: *The Colonial American Jew, 1492–1776. Volume 2.* Jacob Rader Marcus, editor. Detroit: Wayne State University Press. ISBN 0814314031. Pp. 529–547.

When Jews came to the new American colonies between in the 15th and 16th centuries, only a small number were interested in agriculture. Those who bought land or farms usually did it for speculative reasons but made their living in other professions. Among the early Jewish settlers in America, only about 5 percent were involved in agriculture as farmers, ranchers, planters, and viniculture in New England, the Middle Atlantic States and the South.

318. Ornish, Natalie (1989). The Ranchers: Giants of Texas' Great Cattle Herds. In: *Pioneer Jewish Texans: Their Impact on Texas and American History for Four Hundred Years, 1590–1990.* Dallas, TX: Texas Heritage Press. ISBN 0962075507. Pp. 73–84.

This book is a documentary history of the early settlement of Jews in Texas and their involvement in social, political, military, professional, and economic areas of Texas life. Those involved in agriculture were ranchers, cowboys, and landowners with wide influence on the Texas cattle industry. Some combined ranching with trade and shipping ventures. Among the families highlighted here are Mayer and Solomon Halff whose lands covered over 6 million acres on the

Chisholm trail; the Weil Ranches in Corpus Christi; Michael Seeligson who bought 4,442 acres near Goliad, Texas, after serving a few years as mayor of Galveston; Frederick Mayer, who raised cattle and sheep on 85,000 acres near San Angelo; Henry Levy who had 2,140 acres near Victoria and earned the title "Dean of West Texas Ranchmen;" and A. Levytansky who farmed 3,300 acres in LaSalle and Dimmitt Counties. Perry Kallison had a radio program called "Farm and Ranch News" which served the farmers and ranchers of Southwest Texas for forty five years. It became the longest running program of its kind in the nation.

319. Pettus, Louise, and Ron Chepesiuk (1991). Happyville, South Carolina. In: *The Palmetto State: Stories from the Making of South Carolina.* Orangeburg, South Carolina: Sandlapper Publications. Pp. 106–109.

Story of Happyville, a Jewish agricultural colony established in 1905 near Montmorenci, in Aiken County South Carolina. Charles Weintraub, a socialist from New York, acquired a tract of land which he believed would be a suitable site for Russian immigrant families to engage in farming. Weintraub persuaded 25 Russian immigrant families from New York City to settle at Happyville and raise fruits, vegetables, poultry, and livestock. They built houses, cleared land, purchased equipment, built a cotton gin and grist mill and began farming. The colony faced financial difficulties, competition from other farmers, unexpected bad weather, and poor soil conditions. The land was sold in 1908 and the residents returned to New York and New Jersey, some remained in the South and entered non-agricultural jobs. Weintraub left South Carolina and settled in Atlanta.

320. Raisin, Max (1923). American Continent: B. the Russian Period. In: *A History of the Jews in Modern Times.* New York: Hebrew Publishing Company. Pp. 293–301.

Overview of the situation of Russian Jews in America and the influence of the colonization movement. Explains their migration to America and how the lure of agriculture became an attractive alternative to urban life. Contains a brief summary of some of the major settlements and explanations of why most of them failed.

321. Reizenstein, Milton (1906). Agricultural Colonies in the United States. In: *The Jewish Encyclopedia: A Descriptive Record of the History, Literature, Religion, and Customs of the Jewish People from the Earliest Times to the Present Day.* Isadore Singer, editor. New York: Funk & Wagnalls. Pp. 256–262.

An overview of the Jewish agricultural colonies in the United States during the 19th Century. Covers Sholom Colony at Wawarsing, NY, founded in 1837 by 13 Jewish families. The colonies established by the migration of Russian Jews beginning in the 1880s include the Sicily Island, Louisiana; Cremieux and Bethlehem Yehuda, South Dakota; Cotopaxi, Colorado; New Odessa, Oregon; Painted Woods, North Dakota; Lasker and Beersheba, Kansas; Palestine, Michigan; Washington, near Washington, DC; Waterview, Virginia; Chesterfield, Colchester and Montville, Connecticut; and Alliance, Rosenhayn, Carmel, Woodbine, Montefiore, May's Landing, Halberton, Malaga and Hightstown, New Jersey. (See Also: Davidson, Gabriel. Colonies, Agricultural III. United States. In *The Universal Jewish Encyclopedia.* New York, KTAV, 1941. vol. 3, pp. 294–297. For coverage of 20th century agricultural colonies see: *Encyclopedia Judaica.* Jerusalem, Keter, 1971. Pp. 411–413.)

322. Sarna, Jonathan D. (1981). An Asylum for the Jews. In: *Jacksonian Jew: The Two World's of Mordechai Noah.* Jonathan D. Sarna, editor. New York: Holmes and Meier. ISBN 0841905673. Pp. 61–75.

Mordechai Noah (1785–1851) was a playwright, politician, and newspaper editor who attempted to establish a refuge for Jews in Grand Island on the Niagara River in New York State near Buffalo which would be called *Ararat.* He envisioned a temporary self governed colony of agricultural and mechanical arts where new immigrants can live and work and become modernized and acclimated to life in America. He promoted the project through newspapers and public forums and sought widespread support from Jewish leaders, philanthropists and intellectuals in America and Europe. The plan was widely criticized and Noah was subjected to public ridicule. There was very little development on Grand Island and no Jews settled there. Noah abandoned the plan, Grand Island deteriorated and was sold for logging in 1833. In 1852 it was incorporated as a town and today has a population of over 18,000 people.

323. Shulman, Abraham (1976). Agricultural Settlements. In: *The New Country.* New York: Scribner's Sons. ISBN 0684147041. Pp. 176–181.

This book is a collection of captioned photographs and a 52 page introduction depicting life among newly arrived Jewish immigrants who came to America between 1890 and 1924. The photos show typical days at home, work, and religious functions. "Agricultural Settlements" depicts life among immigrants in farming communities in North Dakota, South Dakota, Kansas, Missouri, and Oklahoma.

324. Terwilliger, Katharine T. (1977). The Jewish Colony at Sholem. In: *Wawarsing, Where the Streams Wind: Historical Glimpses of the Town.* Ellenville, NY: Roundout Valley Publishing Company. Pp. 169–175.

This chapter discusses the families involved in the establishment of the Sholem colony and the reasons for their participation. Twelve villagers arrived in Wawarsing, New York 1838 and contracted to build homes and buildings. They erected a synagogue, a store, a museum, an art building, and two factories. One factory made quill pens, the other made caps. The villagers owned their own homes and shared the public building and land. Terwilliger identifies the individuals, their life in the village and the disillusionment that led to its abandonment. Gives details regarding the costs of land and houses, debts and mortgages, and reasons for the colony's failure.

325. Toury, Jacob (1975). M.E. Levy's Plan for a Jewish Colony in Florida —1825. In: *Michael: On the History of the Jews in Diaspora.* Lloyd P. Gartner, editor. Tel Aviv: Diaspora Research Institute. Pp. 23–33.

Moses Elias Levy was born in Morocco in 1782 to a Portuguese father and Anglo-Jewish mother. He was educated in England and later made his way to Cuba where he made his fortune in the lumber business. He began buying lands in Florida in 1818 and moved to the United States where he became a citizen in 1823. Levy wanted to establish a Jewish agricultural settlement in Florida based on adherence to religious law which would emancipate Jews and allow them to have a settlement of their own in a pluralistic nation. Levy made many contacts with business people and friends hoping to gain financial support for his idea. This chapter includes a letter to Sir Isaac L. Goldskin, co-founder of the University College, London and an activist in the struggle for Jewish emancipation in Europe. In the letter, Levy states his reasons for purchasing the lands and his plan for the colony.

326. Uchill, Ida L. (1957). The Yiddish Speaking Jews. In: *Pioneers, Peddlers, and Tsadikim.* Denver: Sage Books. Pp. 165–187. (Reissued in paperback as: *Pioneers, Peddlers and Tsadikim: The Story of the Jews in Colorado.* Boulder: University Press of Colorado, 2000. ISBN 0870815938.)

Describes the Eastern European Jews from Austria, Hungary, Romania, Poland, and Russia came to Colorado during the migrations of the 1880s. Focuses of the Russian Jews who came under the sponsorship of the Jewish Agricultural Society to establish an agricultural colony at Cotopaxi. Gives details about the formation of the colony, the philosophy of members who believed in the communal approach, and some of its active members. The colony was beset from the start with economic problems, religious and political dissension and lack of farming experience that led to its demise in two years. Although the colony was abandoned, nearly all of the members remained in Colorado and entered farming, stock raising, and related fields.

327. Watters, Leon L. (1952). The Bamberger Brothers. In: *The Pioneer Jews of Utah.* New York, American Jewish Historical Society. Pp. 163–169.

Simon Bamberger (1846–1926), was the first non–Mormon, the first Democrat and the only Jew to be elected governor of Utah in 1916. Prior to his election, he supported the agricultural colony at Clarion and offered financial backing. Bamberger convinced prominent Utah citizens to form and manage a Utah colonization fund to help the colony survive. The colony suffered financial difficulties and was out of business by 1915.

328. Weller, George (1937). The Promised Land. In: *New Letters in America.* Horace Gregory and Eleanor Clark, editors. New York: Norton. Pp. 210–219.

Personal observations of 100 Russian and Central European immigrants living and working in the New Jersey agricultural settlements. Weller visited many of the homesteads in those communities and observed their daily lives and spoke with the families. His "letter" offers anecdotes about their lives and economic, political, and sociological perspectives.

Conferences, Symposia, Reunions

329. Alliance (NJ). Golden Jubilee and Fiftieth Anniversary Reunion. (1932 August) *Yōvāl: A Symposium upon the First Fifty Years of the Jewish Farming Colonies of Alliance, Norma and Brotmanville, New Jersey.* Published in Conjunction with the Golden Jubilee and Fiftieth Anniversary Reunion. Philadelphia, The Committee on Arrangements. 47 pp.

In 1882, Jewish refugees from Russia established agricultural settlements in Alliance, Norma and Brotmanville, New Jersey and began life as farmers. This publication is a souvenir journal issued for a symposium in 1932 documenting the history of the colonies and lives of the settlers. Articles written by residents of the three towns and accompanying illustrations present the history and reminiscences of life on the farms, synagogues and religious events, schools, businesses, relations with non–Jewish neighbors, activities of pioneer women, and rosters of the early settlers of Alliance.

330. Benjamin, Eugene S. (1907). The Baron De Hirsch Fund. In: *Proceedings of the Fourth Biennial Session of the National Conference of Jew-*

ish Charities in the United States. Philadelphia, May 6–8, 1906. New York, Stettiner Bros., pp. 156–170.

A paper presented by Eugene S. Benjamin, President of the Baron de Hirsch Fund, stated that the purpose of the Fund "was to assist immigrants and establish them as useful members of the community in which they settled." The Fund does not promote or assist immigration, it deals with the immigrants after they arrive. Benjamin gives details of the five areas of work carried out by the Fund and the relevant facts and costs involved in each area. These are: general aid to the immigrant, removal work, educational work, the Woodbine colony, and industrial activities.

331. Bressler, David M. (1910). The Removal Work, Including Galveston. In: *Proceedings of the Sixth Biennial Session of the National Conference of Jewish Charities in the United States.* St. Louis, May 17–19, 1910. Baltimore, Kohn & Pollock, Inc., pp. 111–166.

The Industrial Removal Office (IRO) was established in 1901 to facilitate the distribution of Jewish immigrants throughout the country. IRO undertook the education, training, housing, and relocation to areas where the immigrants can become a part of the community. Between 1901 and 1909, 45,711 persons were sent to 1,278 cities and towns in the United States and Canada. Bressler explains the activities of the IRO and gives statistics on the number of settlers in each location and their occupations. To illustrate IRO's success, Bressler focuses on the Galveston (Texas) project which sent 2,000 immigrants to settle in that city. The presentation included a brief coverage of the immigrants from Russia and Galicia who were relocated to farming areas in various states. The Jewish Agricultural and Industrial Aid Society supported about 3,000 farmers and their families in over 2,700 farms representing about 50 percent of the actual number of Jewish farmers. Bressler believes Jewish farmers are making an important contribution to American life and he recommends structured agricultural training and education to instill a love of farming and rural life in their children. Following the presentation of Bressler's paper, a discussion took place on how to encourage Jewish immigrants to pursue a life in agriculture. The participants suggested training farms, agricultural schools and free tracts of land as incentives to the newcomers.

332. Davidson, Gabriel (1922). The Tragedy of Sholem. In: *30th Annual Meeting of the American Jewish Historical Society.* New York, April 30, 1922. Philadelphia, The Society. 8 pp.

In this address presented at the annual meeting of the American Jewish Historical Society, Davidson recounts the attempts to establish an agricultural colony called Sholem in the town of Wawarsing, New York. He describes the identity of the settlers, the land costs, economics and conditions in the colony. Sholem failed after a few years because the land was not suited to farming, the growing season was short and when local industries began closing, there were few opportunities for settlers to find off-farm work in the area to supplement their income. Following a series of foreclosures, the colony was fully disbanded by the late 1840s. Davidson interviewed five descendants of the Sholem colony seeking more information about their motivations for joining, what life was like in the colony, and where they went after leaving. (Also published in the *Jewish Tribune*, June 16 and 23, 1922 and Davidson, Gabriel [1943]. *Our Jewish Farmers and the Story of the Jewish Agricultural Society*. New York, L.B. Fischer. Pp. 196–204.)

333. Dubrovsky, Gertrude W. (1978). The Rural Experience in Farmingdale, New Jersey. In: *New Jersey's Ethnic Heritage: Papers Presented at the Eighth Annual New Jersey History Symposium, December 4, 1976*. Paul A Stellhorn, editor. Trenton, NJ, December 4, 1976. Trenton, New Jersey Historical Commission. Pp. 36–58.

A history of the community of Jewish poultry farmers in Farmingdale, New Jersey from its inception in 1919 through the 1950s. The community was settled by four waves of immigrants from Eastern Europe in the early 1900s, from Germany in the 1930s, American born professionals and intellectuals in the 1940s, and Holocaust survivors in the 1950s. Dubrovsky offers personal anecdotes from settlers discussing their reasons for settling in a farm community. She describes life in the community for nearly five decades and how they handled issues like anti-Semitism, political differences, attitudes toward education and religion, the anti-communist frenzy of the McCarthy era, and the economic hardships that often occurred. The community began to decline in the 1950s due to two major factors — political differences which divided the community into factions and competition from large scale agriculture which made it difficult to remain in business. Many farmers sold their farms and returned to manual trades and professions in New York and nearby communities. A few remained in the community and continued to farm, but the structure of Farmingdale as a center of Jewish farming ended.

334. Federation of Jewish Farmers of America (1909–1916). *Programme of the Annual Convention of the Federation of Jewish Farmers of America*. New York, Federation of Jewish Farmers. Pages vary.

Programs in Yiddish and English of the Federation's annual conventions. The programs list participating organizations, exhibits, discussion sessions and speakers, goals and activities of the Federation, the constitution, membership benefits, greetings, and advertisements.

335. Krauskopf, Joseph (1907). Agricultural Education — its Possibilities in Preventive Charity. In: *Proceedings of the Fourth Biennial Session of the National Conference of Jewish Charities in the United States.* Philadelphia, May 6–8 1906. New York, Stettiner Bros. Pp. 195–203.

Rabbi Joseph Krauskopf, President of the National Farm School, explains factors that led to the failure of some of the agricultural settlements. Lack of experience in agriculture, difficulties of draining swamps, hardships of home building and land preparation seemed burdensome for the settlers. In an attempt to alleviate these conditions, two agricultural schools are available to train Jewish youngsters in practical and scientific agriculture. One is the Baron De Hirsch Agricultural School at Woodbine. Krauskopf responds to critics of the agricultural colonization programs with recommendations for better methods to select land and settlers and more focused philanthropy by the sponsoring organizations. He wants to expand agricultural education to include urban youth who may want to leave the crowded ghettos and pursue a livelihood in agriculture.

336. Leucht, I. L. (1907). Possibilities for Agricultural Settlement in the South. In: *Proceedings of the Fourth Biennial Session of the National Conference of Jewish Charities in the United States.* Philadelphia, May 6–8, 1906. New York, Stettiner Bros. Pp. 189–195.

Leucht recommends settling Russian immigrants in the southern states. He points to the rapid colonization by other ethnic groups in the South including German, Italians, French, English, Irish, Hungarians, and Scandinavians. Leucht points to the bright skies, healthy climate, friendliness of the people, good farming conditions, and reasonable land values favorable to southern colonization. Southern agriculture offers year round pastures, thoroughbred horses, high quality beef animals, fecund sheep, fowl of all breeds and a thriving egg industry. Horticulture and truck farming are also being developed. Cotton, the South's major agricultural product, continues to rise in value and production. The South has vast mineral resources, and a growing industrial base of coal, steel, and iron. Leucht asserts that "Russian immigrants should be located throughout the Southern land, in order to become self sustaining, and for the purpose of lightening the great burden of our northern cities."

337. Levy, Abraham R. (1907). Agriculture, a Most Effective Means in Adjusting the Compromised Economic Conditions of the Jewish Poor. In: *Proceedings of the Fourth Biennial Session of the National Conference of Jewish Charities in the United States.* Philadelphia, May 6–8, 1906. New York, Stettiner Bros. Pp. 181–188.

Referring to immigrants from Russia, Galicia, and Romania specifically, Rabbi Levy argues that life on the farm will lift them out of poverty and enable them to lead happy and productive lives. He outlines several advantages of farm life such as hard work which makes them physically fit, obtaining the highest market price for their produce, and being in touch with nature, fresh air, and wholesome food. Levy asserts that living in rural communities has a tendency to strengthen the ties that bind family members and prevent the alienation often found among the urban poor.

338. New Jersey Agricultural Experiment Station. *Lipman Hall, Being an Account of the Proceedings at the Dedication on June 12th 1951, of a Building Named in Honor of Jacob Goodale Lipman. Together with a Review of Fifty Years of Progress in Three Departments Which Grew up under His Guidance at the New Jersey Agricultural Experiment Station, Rutgers University, the State University of New Jersey.* June 12, 1951. New Brunswick, NJ, Rutgers University. 107 pp.

Proceedings at the dedication of Lipman Hall at the New Jersey Agricultural Experiment Station on June 12, 1951. Contains features and statistics about Lipman Hall followed by addresses, dedications, and historical accounts of Lipman's service and renown during his 39 years at Rutgers. Covers Lipman's life, his research, his contributions to agriculture, his years as editor of *Soil Science*, his impact on Rutgers and New Jersey and his overall contributions to American agriculture.

339. Proctor, Samuel (1956). Pioneer Jewish Settlement in Florida 1765–1900. In: *Proceedings of the Conference on the Writing of Regional History in the South with Special Emphasis on Religious and Cultural Groups.* Miami Beach, FL, February 15, 1956. Jewish Theological Seminary of America, New York. Pp. 81–115.

Accounts of some of the Jewish pioneers who settled in Florida during the territorial period following Spanish rule. Includes a detailed biography of Moses Elias Levy, a landowner and politician, who was involved in progressive activities particularly in the field of education. Levy opposed slavery, bigotry, and intolerance and was a spokesman for Jewish causes. He owned one of Florida's largest sugar cane plantations at Micanopy in Alachua County. He attempted to establish a Jewish agricultural colony on the plantation to serve as a refuge for Jews immigrating from Europe but was unable to find suitable economic backing and abandoned the project.

340. Rich, Adolph W. (1907). Plan of Agricultural Settlement. In: *Proceedings of the Fourth Biennial Session of the National Conference of Jew-*

ish Charities in the United States. Philadelphia, May 6–8 1906. New York, Stettiner Bros. Pp. 170–181.

A review of the formation, progress, and anticipated outlook for the agricultural settlement at Arpin, Wisconsin which was established in November 1904 with eighteen families. Rich describes how the families were chosen, the land, buildings, and weather conditions and the requirements for each family including the probation feature that allows for the removal of undesirable settlers. In this paper, Rich gives accounts of life on the settlement after 1½ years. Despite social conflicts, most of the settlers indicated that they were content with the settlement idea. Includes text of the "Milwaukee Plan" and its special features adopted by the Milwaukee Agricultural Association which was involved in the founding of Arpin.

341. Rudin, James A. (1987). The Rural Crisis: A Jewish Perspective. In: *The National Interreligious Conference on Rural Life.* Chicago, February 8–9, 1987. 12 pp.

In the wake of the farm crisis of the 1980s, which resulted in the foreclosure and loss of many small and moderate farms, hate groups have been exploiting the crisis by accusing Jews of a conspiracy to destroy small scale agriculture. This interfaith conference was held to address those issues and find a way to combat the farmers' fears and fight the anti-Semitism. In this paper, Rudin declares that agriculture has always been a vital element of Jewish history and is deeply embedded in Jewish tradition from biblical times to the present. Rudin is concerned about the destruction of the family farm and urges an interfaith initiative among Jewish and Christian communities to take action to confront the farm crisis and seek real and workable solutions that will help farmers avoid foreclosure and counter the anti-Jewish propaganda. (Available online at: http://www.ajc archives.org/AJCArchive/DigitalArchive.aspx.)

342. Sabsovich, S. L. (1907). Agricultural Education. In: *Proceedings of the Fourth Biennial Session of the National Conference of Jewish Charities in the United States.* Philadelphia, May 6–8, 1906. New York, Stettiner Bros. Pp. 203–213.

Dr. Sabsovich, General Agent of the Baron de Hirsch Fund, points out the urgency of encouraging Jewish youngsters to enter agriculture. He discusses the state of agriculture in the United States and the efforts to disseminate information to farmers through the Department of Agriculture and state experiment stations and farmers' Institutes. In addition, Sabsovich explains that the secondary schools and colleges of agriculture around the country are evidence of the need to train American youth for a successful life as farmers. He claims that these schools meet the needs of American agriculture but are not geared to creating Jewish farmers. He calls for the establishment of Jewish agricultural schools that train the largest possible number of Jewish youth in practical farming methods. He also recommends introducing agriculture into orphan asylums and establishing agricultural homes for Jewish delinquent and dependent children.

Dissertations, Theses, Essays

343. Ashkenazi, Elliott (1983). *Creoles of Jerusalem: Jewish Businessmen in Louisiana, 1840–1875.* PhD Dissertation, George Washington University (Washington, DC). 337 pp.

Jewish immigrants who came to Louisiana from Europe and other parts of the South during the early 19th Century were attracted by the opportunity to succeed in business and set up their households in a thinly populated area in rural Louisiana. Ashkenazi explores the role of Jews as businessmen, merchants, farmers, plantation owners, and traders during the 35 year period from 1840–1875. The State's economy during those years comprised a large agricultural sector and a commercial center in New Orleans. The agricultural sector relied heavily upon entrepreneurs in New Orleans who marketed their crops and sold them supplies. Among the prominent Jewish individuals in agriculture were Jacob Lemann, a merchant and owner of several sugar plantations, Abraham Levi and Henry Lehmann, cotton traders, and Joseph Embree, a successful farmer. This study describes the commercial activities of Louisiana Jews, the centrality of agriculture in the State's rural economy, and the social and economic interactions among Jewish and non–Jewish neighbors.

344. Bentley, Melissa (2003). *The Van Eeden Settlement: Alvin Johnson's Attempt to Rescue Jewish Refugees and Turn the Tide of American Public Opinion in Favor of Jewish Resettlement in the 1930s and 1940s.* Senior Thesis in Bachelor of Arts, University of North Carolina at Asheville (Asheville, NC). 23 pp.

Van Eeden was an agricultural settlement established in 1939 on one thousand acres in Pender County, North Carolina. It was planned as a settlement for refugees from Nazi Germany to begin life in the United States in an agricultural setting. Alvin Johnson, Director of the New School for Social Research in New York, established a corporation to fund the settlement project. The settlement was designed as a market gardening business financed by the corporation which expected a return on its investment. The settlers, a diverse group who had lived middle to upper class lives in Germany, were initially excited by the idea but were unaware of what was expected in such a community. Demands of farm life, lack of agricultural experience, long hours of labor, quarrels among members, envy, distrust, and rivalries all played a role in the collapse and liquidation of Van Eeden in 1943. Bentley gives a fascinating account of the Van Eeden settlement. She explains what motivated Johnson to establish the colony, skepticism about its success, views of the refugees, and the factors that doomed the settlement. (Also available online at: http://toto.lib.unca.edu/sr_papers/srhistory_2003/bentley,%20melissa.doc. For more information on the founding and early years of Van Eeden see: Johnson, Alvin A. Van Eden: A Jewish farm community. A vivid picture of an experiment in human reconstruction. *American Hebrew*, 147[25]:5, 17, November 1, 1940.)

345. Blumenthal, Helen E. (1975). *New Odessa: United We Stand—Divided We Fall, 1882–1887.* M.A. Thesis, Portland State University (Portland, OR). 102 pp.

A history of the agricultural colony of New Odessa established in 1882 on 180 acres in Southern Oregon. The settlers were a small group of 65 Russian immigrants that later grew to 90 who had been members of Am Olam in Russia and were committed to living a communal life at New Odessa. Although they tried to make the colony a thriving settlement, they encountered a host of problems which included unfavorable land and soil conditions, lack of agricultural training, political discord, privacy concerns, and other complications that led to the demise of the colony in 1887. Blumenthal describes the anti-Semitism and poverty in 19th Century Russia which sparked their emigration to America. She examines the socialist philosophy of the Am Olam movement and the ideology that attracted the settlers to the New Odessa experiment. She analyzes the political, social, and economic forces that determine the success or failure of utopian communities.

346. DeFeo, Lawrence Joseph (1979). *The Transformation of a Community: A History of the Development and Education of Woodbine.* DEd (Doctor of Education) Dissertation, Rutgers, The State University of New Jersey (New Brunswick, NJ). 269 pp.

A history of the educational activities at the Woodbine agricultural colony based upon interviews and communications with former residents and persons involved with the school, and newspaper accounts. DeFeo sets the stage with the historical context of imperialist Russia and Eastern Europe in the 1800s which sparked the great migration and the emerging social movements in America during that period which he calls the "Progressive Era." He focuses on the establishment of an educational system at Woodbine agricultural colony that was designed to educate the children from grades one through eleven and the Baron de Hirsch Agricultural and Industrial School designed to teach agriculture to the residents of the colony. The schools emphasized a progressive education in accordance with the theories of John Dewey and other educators of that era and played a role as the major influence on acclimating the immigrants and their children to American life. This comprehensive analysis of Woodbine's education system includes details about the buildings, courses, students, costs, and personal experiences.

347. Douglas, Martin I. (1960). *Chronological Summary of Annotated Cards Toward the History of the Jewish Agricultural Colonies in South Jersey.* PhD dissertation, Jewish Theological Seminary of America (New York, NY). 342 pp.

A chronological history of the agricultural colonies in Southern New Jersey from 1882 to 1919 focusing on Vineland, Rosenhayn, Carmel, Alliance, and Woodbine. In this intriguing account of life in the colonies, readers get a look at the daily events and activities that shaped the colonists' lives and helped them survive. Douglas describes the diversity of settlers, their needs, activities, and attitudes towards farming and the community. The settlers provided education for their children, built synagogues, created communal organizations, supported Zionist causes, and led Jewish lives. To supplement their income, factories were established in Vineland, an agricultural school was created in Woodbine, and Rosenhayn became a summer resort. Settlers discovered early that lack of agricultural experience, hard work, and weak cooperation from sponsoring agencies hampered the colonies in the early years. Political conflicts, debts, anti–Semitism, the failure to convince their children to remain on the farms, and the rapid changes in agriculture following WWI led to the phasing out of the communal settlement concepts and transition to individual ownership. (See also: Brandes, Joseph and Martin Douglas. *Immigrants to Freedom: Jewish Communities in Rural New Jersey Since 1882.* Philadelphia, University of Pennsylvania Press, 1971).

348. Eisenberg, Ellen (1990). *Jewish Agricultural Colonies in Southern New Jersey: The Processes of Migration, Settlement and Adaptation.* University of Pennsylvania (Philadelphia, PA). 377 pp.

Eisenberg examines the patterns of migration and adaptation and the social, religious, and cultural aspects of Russian and Polish immigrants in the agricultural colonies of Salem and Cumberland Counties, New Jersey from 1880 through 1920. Focusing on Alliance, Norma, and Brotmanville in Salem County and Rosenhayn and Carmel in Cumberland County, she analyzes the differing roots and life styles prior to immigration and its impact on their adaptation to communal agricultural colonies. She contrasts the socialist and agrarian perspectives of the immigrants with Western capitalist ideas and how these differing perspectives shaped the character of the colonies. (See also: Eisenberg, Ellen. *Jewish Agricultural Colonies in New Jersey, 1882–1920.* Syracuse, Syracuse University Press, 1995. ISBN 0815626525.)

349. Goldstein, Philip Reuben (1921). *Social Aspects of the Jewish Colonies of South Jersey.* PhD Dissertation, University of Pennsylvania (Philadelphia, PA). 74 pp.

An analysis of the agricultural colonies in southern New Jersey. Goldstein introduces the topic with a discussion of the status of Jews and the rampant anti-Semitism in Russia that sparked a wave of immigration to the United States. He describes the founding of the colonies of Rosenhayn, Carmel, Garton Road, Brotmanville/Alliance and Woodbine and the personalities and organizations that offered support. Goldstein examines the population, economics, transportation and communication, and businesses in the mixed agro-industrial settlements. By 1919 younger residents were leaving the colonies to pursue lives in urban and suburban areas in non-agricultural professions. He explores the reasons for their lack of interest in remaining on the farm and offers some suggestions on revitalizing the settlements to help them survive.

350. Gordon, Morton L. (1974). *The History of the Jewish Farmer in Eastern Connecticut.* PhD Dissertation, Yeshiva University (New York, NY). 244 pp.

This dissertation presents the agricultural history of Jews in Connecticut that began with a small number of immigrants from Russia and Germany and grew as more immigrants came to rural Connecticut seeking a life on the farm. Settlements arose in Chesterfield, Uncasville, Norwich, Colchester, Lebanon, Ninantic and Modus and the fertile Connecticut River Valley near Hartford, Wellington, Sobers, and Rockville. Major growth and expansion occurred during two waves of immigration from 1933–1954 into successful poultry, dairy and vegetable enterprises. Gordon discusses the influence of farming on lives of immigrants, relations with non-Jewish neighbors and financial and educational assistance provided by the Jewish Agricultural Society. The farm families were committed to preserving their Jewish heritage by establishing, synagogues, religious schools, community centers, and libraries. The communities began to wane in the 1950s with the trend toward large scale farming. Many of the farmers retired or entered other professions and their children left for better opportunities in the cities. Gordon points out that for a 25 year period, agriculture served as the medium through which thousands of immigrants found new life and hope in the United States.

351. Harris, Lloyd David (1984). *Sod Jerusalems: Jewish Agricultural Communities in Frontier Kansas.* M.S. Thesis, University of Oklahoma (Norman, OK). 190 pp.

A study of the Jewish agricultural colonies established in Western Kansas during 1882–1886. Seven colonies were established for newly arrived Russian immigrants who were seeking alternatives to life in urban areas. The Beersheba, Montefiore, Lasker, Hebron, Gilead, Touro, and Lesser colonies lasted a short time and most were gone by 1900. Each colony was unique in regard to how it was established, the secular and religious lives of its inhabitants, and the farming practices. Harris explains how the colonies were established and the factors contributing to the failures. (Available Online in Full Text at: Http://www.Kancoll.org/books/harris/index.html.)

352. Katz, Nancy Harris (1964). *Jewish Agricultural Colonies in the United States.* M.A. Thesis, Southern Methodist University (Dallas, TX). 54 pp.

A historical view of the Jewish agricultural colonies that were established in the 1880s and the following decades through the 1940s. Katz begins by analyzing the mass emigration from Russia and the attitudes of the immigrants and relief agencies regarding the establishment of agricultural colonies. She briefly describes settlement activities prior to 1880 to show that Jews were engaged in agriculture from early times. She provides a comprehensive study of colonies formed in the 1880s in Louisiana, Arkansas, Colorado, the Dakotas, Kansas, and Oregon. Katz treats the New Jersey colonies of Alliance, Rosenhayn, Norma, Brotmanville, and Woodbine separately because they survived for a longer period and served as prototypes for what can be achieved by well designed and managed communal settlements. She points out that although many of the 1880 colonies were not successful, new colonies and group settlements continued to be formed from the 1890s through the 1940s in New England, the Midwest and Northern California and homesteads established during the depression years in New Jersey, Michigan, and other states. Although settlers were faced with economic problems, unsuitable land and soil conditions, inclement weather, political conflicts, and competition from large scale farming, Katz explains that each settlement and community had a unique history and characteristics that determined their successes or failure. The appendix contains a chronological listing of the founding and ending dates and number of families of 44 colonies in 20 states.

353. Kirk, Marvin Steward (1926). *A Study of the Jews' Contribution to Land Settlement and Land Credit with Special Reference to North Dakota.* MS Thesis, North Dakota State College of Agriculture and Mechanic Arts (Fargo, ND). 54 pp.

Kirk gives a history of Jewish settlement in Palestine, Canada, and the United States and then focuses on the North Dakota experiences. He describes the settlement movements, the organizations and philanthropists that supported the settlement concepts, and the costs, loans, mortgages and other economic considerations that impacted the settlements. Jewish agricultural settlement in North Dakota began in 1887. In 1925 there were 890 settlers in colonies or individual farms in 9 North Dakota counties who were granted over $517,000 in loans by the Jewish Agricultural Society. The settlers were sent to the colonies with no agricultural knowledge or experience and insufficient means of support to learn farming techniques or to purchase equipment and supplies. Most of the colonies were located in places which were not convenient to market, had poor soil, were subject to drought, hails storms, prairie fires, and high rates of interest. The colonies failed within 3–5 years as settlers left to enter trades or businesses in nearby towns and villages or other states.

354. Lamb, Blaine Peterson (1982). *Jewish Pioneers in Arizona, 1850–1920.* PhD Dissertation, Arizona State University (Tucson, AZ). 385 pp.

Historical account of Jews in Arizona from the 1850s through 1920. Lamb covers in detail the contributions of Jews to the economic, social, and political advancement of early Arizona. Arizona Jews were involved in merchandising, storekeeping, mining, agriculture and banking. Chapter VI (pp. 179–213) describes the active participation of Jews as owners and operators of ranches, promoters of irrigation projects, milling enterprises, and distribution and marketing of agricultural products. Lamb traces the lives of some prominent Jewish personalities in Arizona agriculture and presents their backgrounds, how they got involved in farming, and the areas where they lived and worked. Many merchants also ran livestock on nearby pastures, leased irrigated farmland, and sold grain, beef or wool. Others supplied farmers and ranchers with feed, seed, fertilizer, and manufactured goods. During the late 19th century, many farmers and ranchers sold large portions of their crops and livestock to government contractors to supply Federal institutions in the territories. The activities of the Jewish merchant-contractors, farmers and ranchers played a major role in the development of agriculture in Arizona.

355. Lazar, Robert Jordan (1968). *From Ethnic Minority to Socio-Economic Elite: A Study of the Jewish Community of Fargo, North Dakota.* PhD Dissertation, University of Minnesota (Minneapolis, MN). 329 pp.

A study of the history the Jewish community of Fargo, North Dakota and the social and cultural factors that led to settlement in the area. Data was gathered from questionnaires, interviews, records from Fargo Hebrew Congregation, Temple Beth El and other sources covering 131 families in Fargo and Moorehead, Minnesota having at least one Jew present. The study traces the European backgrounds of the families, and the historical experiences from first arrivals to the present and their status in the community. Although the Jews of Fargo were not engaged in agriculture, the author sets the framework for Jewish settlement in North Dakota in Chapter II "The Jews of North Dakota—Early Agricultural Colonies" (pp. 39–67) by presenting an overview of Jewish settlement in agricultural colonies in at Devils Lake, Painted Woods, Grand Forks and Iola in the early 1900s.

356. Levine, Darwin Solomon (1939). *A Brief Survey of the Activities of Jews in American Agriculture.* M.A. Thesis, Columbia University (New York, NY). 147 pp.

Levine presents a chronological history of Jews in Agriculture from 1492 to 1927. The information was gleaned mainly from articles in Jewish newspapers such as *American Hebrew, American Israelite, Jewish Tribune* and some reports and archival documents. Part I—1492–1880 highlights sporadic attempts by individual Jews to enter farming in North and South America beginning with Luis de Torres who came to the New World with Columbus. Accounts of Jews in agriculture in North America began in the 1650s. Part II—1880–1927 was the period of organized Jewish agricultural settlement by Russian and German Jewish immigrants. Levine gives descriptions of settlements in Maryland, Georgia, Virginia, Florida, North Dakota, South Dakota, Colorado, Michigan, Wisconsin, Utah, Massachusetts and New Jersey. Part III—Examines some of the factors that made agriculture an attractive alternative to urban living for new immigrants and factors that led to their success or failure.

357. Naftaly, Phillip R. (1999). *Jewish Chicken Farmers in the Egg Basket of the World: The Creation of Cultural Identity in Petaluma, California, 1904–1954.* PhD Dissertation, New School for Social Research (New York, NY). 267 pp.

A study of Jewish immigrants who came to Petaluma, California in the early 1900s and established a community of chicken farmers. The community flourished for 50 years as a center of Yiddish culture motivated by socialist ideologies. Using data gathered from historical research and interviews with retired Jewish chicken farmers, Naftaly examined the development of the community and the social and economic aspects that sustained it with particular focus on four factors: historical Jewish experiences in group settings, anti-Semitism, socialist and Zionist ideologies, and the relationships between agriculture, "Americanization" and ethnic identification. The findings indicated that "connections to the broader non-diasporic societies in which ethnic groups reside may be more determinative than connections to a diaspora community, especially with regard to generational change."

358. Pearlstein, Peggy Kronsberg (1993). *Understanding through Education: One Hundred Years of the Jewish Chautauqua Society, 1893–1993.*

PhD Dissertation, George Washington University (Washington, DC). 451 pp.

The Jewish Chautauqua Society was founded in 1893 by Henry Berkowitz, a Reform Rabbi. The Society's mission was to bring a greater knowledge of the Jewish heritage to Jews and to teach non-Jews about Judaism. This dissertation is a study of the society and its programs, education, and outreach which promoted the integration of Judaism into American life. Among its activities was the establishment of religious schools to educate the immigrant children of Jewish farmers in southern New Jersey and the Dakotas. With assistance from the Jewish Agricultural Society and other agencies, Chautauqua-trained teachers established classrooms and learning facilities, provided educational materials and managed the social, religious, and educational activities in the agricultural settlements.

359. Rosenberg, Debbie L. (1999). *Philanthropic Ideals and a Displaced Reality: American Jews, Immigration, and Agriculture, 1880–1920.* M.A. Thesis, Portland State University (Portland, OR). 150 pp.

An examination of the agricultural colonies settled by Russian Jewish immigrants in 1880 and the social, religious and economic forces that led to the end of the colonies by 1920. Rosenberg gives a brief description of American Jewish history and the status of American Jews in the 1880s. She traces the chronological development of settlements that were established for Russian immigrants in Louisiana, New Jersey, Virginia, Colorado, Kansas, North and South Dakota and Arkansas and the commitments of Jewish philanthropists to the settlement efforts. Reform Judaism joined with the philanthropic organizations in attempting to meet the settlers' economic and social needs. Rosenberg examines the contradictions between the socialist perspectives of the Oregon settlers and Orthodox colonists in Kansas and its impact on the Reform movement. She concludes that although the colonies ultimately failed they succeeded in drawing immigrants to the Nation's interior and enabled them to adapt to life in America.

360. Satt, Flora Jane (1950). *The Cotopaxi Colony.* M.A. Thesis, University of Colorado (Boulder, CO). 83 pp.

The story of the Cotopaxi Colony in Fremont County, Colorado which existed from 1882 to 1885. The colony, established by Jewish immigrants from three Russian provinces comprised 63 persons from 22 families. Satt describes the topography and resources of the region which were mainly mining and small scale farming. She gives details about the conditions of the Jews in Russia and the settlement philosophies that motivated them to leave and pursue agricultural careers in the United States. Satt focuses on the settlement program and the economic support and assistance they received from the Hebrew Emigrant Aid Society (HEAS) and the reasons for the colony's failure. Emanuel Saltiel, a prosperous resident of Fremont County who made his fortunes in mining and milling, sold the land to the settlers and agreed to provide housing, equipment, household supplies, and farm animals. Saltiel failed to meet his commitments and the colonists had to rely solely on the efforts of HEAS and Jewish Agricultural Society for their existence. The colony was discontinued in 1886 due to a variety of factors that included unsuitability of the land for agriculture, economic deprivation, discontent among the members, and dwindling support from HEAS and other organizations. Nearly all the members remained in Colorado and continued as independent farmers or entered new professions in business and commerce. (For a response to Satt's coverage of Saltiel's involvement in the failure of Cotopaxi see: Saltiel, Miles. The Jewish Agricultural Colony at Cotopaxi: Rebalancing the Record. Part 1. *Rocky Mountain Jewish Historical Notes* 19(3–4):1–7, Winter/Spring 2005 and Part II *Rocky Mountain Jewish Historical Notes* 20(1, 2):1–11, Fall/Winter 2005.)

361. Shally-Jensen, Michael (1992). *New Deal, New Life: Culture and History of a Jewish Cooperative Colony in New Jersey, 1933–1939.* PhD dissertation, Princeton University (Princeton, NJ). 555 pp.

Jersey Homesteads was a cooperative settlement of about 200 families established by East European Jewish immigrants in 1936 in what is now known as Roosevelt, New Jersey. Jersey Homesteads was planned as a cooperative rural community with a mixed agricultural and industrial revenue base. The colony provided for one hundred twenty garment workers and their families from New York and Philadelphia who lived on the colony and worked in the garment factory. The agricultural wing of Jersey Homesteads began with five couples who were later joined by other homesteaders thus increasing to ninety seven families. Using oral histories, historical documents, various archival files, and newspaper and magazine articles, Shally-Jensen examines the theoretical framework, cultural patterns, major personalities, historical development, and the ever present economic and ideological issues, and internal conflicts that brought both progress and enmity to this utopian experiment. He analyzes the conflict management and the use of language and thought as strategies that influenced the colony's operation. Political, cultural, and philosophical differences led to dissolution of the colony by the end of World War II.

362. Shevlin, Arnold (1950). *A Study of Two Jewish Communities: Rockville, Connecticut and Norwalk, Connecticut.* M.S Thesis, Hebrew Union College (Cincinnati, OH). 64 pp.

This study compares the Jewish population of the small urban town of

Rockville and nearby Ellington and the urban community of Norwalk and Westport, Connecticut. The purpose of the study was to obtain information about the Jewish families and their characteristics in order to organize social and religious activities that fit the communities' needs. A questionnaire was sent to the families to gather data on ages, occupations, political views, economic conditions, social life, and Jewish involvement. In 1948 Rockville had sixty Jewish families and Wellington had fifteen families with about twenty one farm families among them. In the first decade of the 20th Century, the Hebrew Emigrant Aid Society settled Russian immigrants in the area and in 1907 they established the Association of Connecticut Jewish Farmers. Although the farmers were affected by the economic depression of the 1930s, their farms were not repossessed and they remained on the land. At the time of this survey, they were prosperous large scale farms with higher average income than the Jewish population as a whole. The farms produced tobacco and potatoes; some farmers were also involved as food brokers and some had farms in other parts of New England. One of the farmers donated a section of land to Yale University for experimental purposes. The Norwalk/Westport area had approximately seven hundred fifty Jewish families in 1948. None of those families were engaged in farming.

363. Silver, Harold (1934). *Some Attitudes of the East European Jewish Immigrants Toward Organized Jewish Charity in the United States in the Years 1890–1900.* MS Thesis, Graduate School for Jewish Social Work (New York, NY). 329 pp.

A study of the attitudes of East European Jews towards organized charity in the decade 1890–1900 as reflected in the Yiddish Press. Using articles, letters and columns from 12 Yiddish newspapers, Silver reveals attitudes of Russian immigrants toward German Jews and the policies, viewpoints, and programs of the agencies. Claims of hostile treatment and inadequate assistance in issues of housing, labor, and social welfare were commonly reported. Organized labor criticized the agencies for their hostility to unions and for using immigrants as strike breakers. The immigrants were critical of some of the tactics of the Baron De Hirsch fund and the colonization effort. They accused the fund of using the colonization movement to keep them from settling in urban areas, lying about the destinations and type of work, and failing to offer adequate training, education, economic assistance and follow up support for the colonies. The Woodbine Colony received most of the attention in the Yiddish press which reported labor troubles in the factories, autocratic methods in supervision of the farm school, and dispossession of families that had conflicts with management. The Russian Jews began to establish their own institutions and eventually achieved friendly relations and mergers with German Jewish agencies.

364. Singer, Richard E. (1941). *The American Jew in Agriculture: Past History and Present Condition.* Hebrew Union College Prize Essay 1941, Hebrew Union College (Cincinnati, OH). 257 pp.

This comprehensive work traces the history of Jewish involvement in American agriculture in settlements in 35 states and the District of Columbia from the colonial period through 1940. The arrangement is chronological as follows: Colonial period from 1492 to 1775, post revolutionary period 1776–1880, Russian refugee period 1880–1914, the recent period 1914–1940. Provides details and overviews about the people involved, locations where they settled, analysis of the soils and farming conditions, the organizations that supported them, and life on the farm and in their communities. Although the author used a wide range of periodicals, books, government documents and miscellaneous sources, most of the details regarding the settlements came from the newspapers *American Israelite,* and *American Hebrew* and annual reports of the Jewish Agricultural Society.

365. Spiegel, Alfred D. (1939). *History of the Jewish Agricultural Colonies in Southern New Jersey.* Bachelor of Arts Thesis, Moravian College (Bethlehem, PA). 41 pp.

History of the Jewish agricultural colonies of South Jersey established in the 1880s and 1890s. Spiegel begins with a brief history of Jewish life in Russia and the organizations that arose to assist the immigrants in their adaptation to America and a life in agriculture. The remainder of the paper is a detailed study of the six successful colonies of Alliance, Norma (where Spiegel spent his childhood), Brotmanville, Rosenhayn, Carmel, and Woodbine. Spiegel gives details of the acquisition, size and costs of each colony, farming and factory output, life in the colonies, the role of organizations that supported them, and how they managed to survive through the early 20th century.

366. Starkoff, Bernard (1940). *The American Jew in Agriculture: History and Present Condition.* Rabbinical Thesis, Hebrew Union College (Cincinnati, OH). 264 pp.

A description of the development and history of the Jewish agricultural movement seen through seventy seven settlements established between 1837 and 1933. Starkoff looks at the origins, development and successes or failures of settlements in New York, New Jersey, Connecticut, Michigan, and Ohio and the organizations associated with the movement. He examines the depression era of the 1930s and its impact on Jewish farmers and contemporary agriculture. By 1933, the economic condition of the agricultural sector had improved due to a rise in farm commodity prices and increases in consumer buying power as a result of legislation passed by President Roosevelt. This prompted the establishment of additional attempts at Jewish farm settlement. Starkoff describes the settlements, their founders and philosophy, and daily life in the settlements. He focuses heavily on cooperative and economic incentives such as credit unions and crop and fire insurance. Starkoff concluded that agriculture does not offer incentives to attract Jews to farming but those who choose a life in agriculture should be well trained and make provisions for the social, intellectual, and religious needs of the family.

Reports from Government Agencies, Organizations, and Associations

367. Baron de Hirsch Fund (1900–1908). *Report.* New York: Press of Philip Cowen.

The annual reports of the Baron de Hirsch fund cover information about the Fund's philanthropic activities on behalf of Jewish communities around the world. The volume for 1902 contains the 8th annual report of the Baron de Hirsch Agricultural and Industrial School (other reports of the School were issued separately). Gives details of the school's activities, courses, buildings, farm production and sales, economic expenditures, student achievements, the school farm and other features of the past year at the agricultural school.

368. Bloom, Sol (1925). *The Jewish Farm Movement of To-Day: A Quarter Century of Progress by American Jewish Farmers. Speech in the House of Representatives, Monday, March 2, 1925,* in: *Congressional Record. Proceedings and Debates of the Second Session of the Sixty-Eighth Congress of the United States. Vol. 66 Part 5,* pp. 5163–5166.

Bloom presents a history of Jewish agricultural settlement in the United States and the successes and contributions of individual farmers and farming communities. He describes the efforts of the Jewish Agricultural Society to extend financial assistance to help Jewish farmers remain in agriculture and scholarship programs for their children. Bloom asserts that Jews have a role to play in the development of American agriculture.

369. Bressler, David M. (1907). *The Distribution of Jewish Immigrants in Industrial and Agricultural Pursuits; Agencies and Results.* New York: Industrial Removal Office. 32 pp.

The Industrial Removal Office (IRO) was established in 1901 with funds from Baron De Hirsch to organize the distribution of Jewish immigrants, mainly from

Eastern Europe to communities throughout the country. IRO programs were designed to relieve congestion in the cities and give immigrants the opportunities to become self sufficient in new environments. From February 1901 to February 1907 over 30,000 persons were moved from New York, Philadelphia, and Boston to towns and cities in 37 states to engage in new occupations and trades. IRO assisted the Jewish Agricultural Society with relocation to agricultural settlements in New Jersey, New York, Connecticut, the Dakotas, Massachusetts, Florida, Michigan, Oklahoma, and Illinois. The report briefly describes the localities, the settlement patterns, economic support, and the types of enterprises best suited to each settlement area.

370. Clarion Colony, Utah (1912). *First Successful Jewish Colony in the United States.* Salt Lake City: Utah Colonization Fund. 16 pp.

This report was issued by the Utah Colonization Fund to gain publicity for a bond issue to raise funds for establishment of the Clarion Colony in Utah. The idea began with organization of the Jewish Agricultural and Colonial Association in 1910 which signed up 150 members from New York and Philadelphia. The colony was established in 1911 and had its first harvest in 1912. Describes how the land was purchased, developed, settled, and financed with detailed estimates of assets and liabilities. Includes photographs taken at the celebration of the first harvest attended by the Governor of Utah.

371. Davidson, Gabriel (1950). *The Jewish Agricultural Society: Report of the Managing Director for the Period 1900–1949.* New York: Jewish Agricultural Society. 47 pp.

The Jewish Agricultural Society was established in 1900 by Baron Maurice De Hirsch to promote farming among Jews in the United States. The society established agricultural colonies around the country which were settled mainly by new immigrants from Eastern Europe. Some of the settlements were successful and survived for long periods; others were unsuccessful and went out of existence in a few years. This report reviews fifty years of the Society's activities. It begins with a biography of Baron Maurice De Hirsch and his philanthropic ventures. Contains highlights about settlement patterns, financial outlays, development programs, and details about individual farmers with captioned photographs. Covers activities in Connecticut, Massachusetts, North Dakota, New Jersey, Michigan, Ohio, Louisiana, California and other locations.

372. Dillingham, William Paul, and Cance, Alexander E. (1911). Hebrews in Agriculture. In: *Immigrants in Industries. Pt. 24 Recent Immigrants in Agriculture.* Volume 22. Part III, 61st Congress, 2d session, 1909–1910. Serial set no. 5683. Washington, DC: U.S. Government

Printing Office. Pp. 3–146. (Reprinted 1970 Arno Press and New York Times.)

A detailed analysis of "Hebrews" in agricultural pursuits from data supplied by the Jewish Agricultural and Industrial Aid Society of New York and the 1900 census. There was an estimated 3040 Hebrew farmers in 36 states. The report discusses origin of the Hebrew colonies, successful colonization, kinds of agricultural pursuits, standard of living and the Hebrews as citizens. Gives descriptions of specific farming communities in Connecticut, Massachusetts, New York, New Jersey and Wisconsin engaged in general, dairy, tobacco, fruit and vegetables, and summer boarder agriculture. Available online at: http://site.ebrary.com/lib/stanfordimmigrationdillingham/EDF?id=10006605.

373. Duckat, Walter (1955). *Contributions of Jews to the American Economy.* New York: Jewish Occupational Council. 21 pp.

This report, issued to commemorate 300 years of settlement by Jews in the United States, analyzes the contribution of Jews to all areas of the American economy. Duckat names the prominent individuals and their roles in the professions, economics and commerce, small business, manufacturing, trades, transportation, agriculture, entertainment and labor unions. The agricultural contribution includes the Jewish Agricultural Society and the Industrial Removal Office which provided the opportunity for immigrants to settle on farms around the country. Other contributors are a large fruit packer who founded the International Institute of Agriculture, distinguished agricultural scientists and professors whose research focused on bacteriology and soil science, a prominent veterinarian who developed vaccines to prevent poultry diseases, and an organizer of the first forest research unit at the United States Department of Agriculture.

374. Elstein, Leo F. (1892). Sanitary Report as to Russian-Hebrew Colonies in New Jersey. (Prefatory Note by the Secretary New Jersey State Board of Health. In: New Jersey. State Board of Health. *Annual Report of the Board of Health of the State of New Jersey.* Camden, NJ: New Jersey State Board of Health. Pp. 371–383.

Elstein, a Russian physician, visited ten agricultural colonies in southern New Jersey to assess the sanitation and health situation. He visited four colonies established with the assistance of the Jewish Agricultural Society — Carmel, Rosenhayn, Alliance, Woodbine — and six colonies which were independently established by land speculators — Mizpah, Ziontown, Hebron, Malaga, Reega and Albertown. The report covers the history, climate, soil, sanitary conditions, and appearance of each settlement and how well the local boards of health were assisting the settlers in maintaining a clean and sanitary existence. He found a range of conditions from those that are improving and maintaining clean disease free settlements to those struggling to keep going and lacking in proper sanitation. Elstein's descriptions include some biased and stereotypic statements about the settlers.

375. Goldman, Julius (1882). *Report on the Colonization of Russian Refugees in the West.* New York: Evening Post Job Print Office. 35 pp.

Report of a committee charged by the Hebrew Sheltering and Immigrant Aid Society of America to investigate suitable locations for agricultural colonies for new Russian immigrants. The report focuses on four areas of concern: qualifications needed for successful colonization, amount of capital needed to insure success, the most advantageous location in the northwest, and the best approach to carry out the settlement plans. The report outlined the needs of the farm families, the costs and benefits, the suitability of the land for farming, and proposed organization and governance of the colonies.

376. Hebrew Emigrant Aid Society of the United States (1882). *Report of Moritz Ellinger.* New York: Steam Press Office, Industrial School Hebrew Orphan Asylum 1882. 33 pp.

Ellinger traveled to London, Paris, and Frankfort at the request of the Hebrew Emigrant Aid Society (HEAS) to seek the assistance of European Jewish leaders in facilitating the immigration of Russian Jews to the United States and Canada. In this report to the President of HEAS, he gives account of the meetings, the leaders who attended and the financial and logistic support he requested. Ellinger explained that the immigrants will be dispersed to agricultural colonies in the United States where they will be able to lead productive lives. According to Ellinger's account, the European leaders offered their support for the plan and pledged various sums of money to help HEAS absorb the immigrants and carry out the settlement programs.

377. Jewish Agricultural Society (1954). *Jews in American Agriculture: The History of Farming by Jews in the United States.* New York: Jewish Agricultural Society. 62 pp.

A chronological history of Jews in American agriculture in the late 19th and early 20th centuries. The major period of large scale settlement in agricultural colonies occurred in the 1880s with the influx of Jews from Eastern Europe. The report discusses the work of the of the Baron de Hirsch Fund and the Jewish Agricultural Society and its efforts to settle the immigrants in agricultural colonies in New Jersey, New York, Connecticut, Massachusetts, Kansas, North Dakota, California and other locations. It covers the Society's activities, settlement patterns, kind of farming enterprises, progress, and successes and failures of the settlement efforts. Also includes a brief history of Jewish involvement in agriculture during the 300 years of Jewish settlement in America.

378. Jewish Agricultural Society (1933). *The Jewish Agricultural Society: Its Aims and Objectives.* New York: Jewish Agricultural Society. 4 pp.
The society maintains that the family farm is the backbone of American agriculture and its goal is to assist those who want to settle on farms either independently or as part of a cooperative venture. The Society's programs are administered through its departments which include settlement, loan, employment, and rural sanitation departments. A description of the role of each department is presented.

379. Jewish Agricultural Society (1900–1958). *Annual Reports.* New York: Jewish Agricultural Society.
Comprehensive annual reviews of the Society's activities on behalf of Jewish farmers around the country from 1900 to 1958. All aspects of the economic, social, and educational work for the respective years are given. They contain reports by the President and General Manager regarding the settlements, the Society's activities and *The Jewish Farmer* newspaper. The various bureaus and branches of the Society report facts and statistics on loans granted, closed, and outstanding; farm employment; sanitation; secular, religious, and agricultural education, and cooperative ventures in purchasing, credit unions, savings and loan associations, insurance programs, and marketing strategies. (Reports for 1900–1921 issued under the Society's earlier name: Jewish Agricultural and Industrial Aid Society.) Report for 1957 issued with the 50th Anniversary issue of *The Jewish Farmer*, May 1958.

380. Jewish Agriculturalists' Aid Society of America (1900). *Report of the Society's Work and Achievements from the Time of Its Organization, (Fall of 1888) to November 1st, 1900.* Chicago: The Society. 36 pp.
The Jewish Agriculturists' Aid Society of America was established in 1888 to encourage and aid Jewish people to engage in agriculture in any section of the United States and Canada. The Society offered financial assistance and guidance to acquire and improve land and buildings and outlays for implements, machinery and livestock to run their farms. This report covers the activities and accomplishments of the Society in its first 12 years. Includes details of the location and conditions of homesteads and rented farms, number of Jewish farmers in the Midwest, financial report, and list of contributors.

381. Kirsch, William (1920). *The Jew and the Land.* Madison, WI: American Association for Agricultural Legislation (Bulletin No. 7). 46 pp.

This monograph is an attempt to investigate whether Jews are fit for a life in agriculture and to devise a plan to facilitate the move from urban to rural life for those Jews who choose to farm. The author depicts Jews as a people outside of agriculture whose "inherited racial characteristics" and "psychical traits" have adapted them to urban life in professions and businesses that involve a struggle for profits. He outlines the distinct differences between farm life and urban life and asserts that the Jews would have to undergo a period of adaptation to become attuned to an agricultural environment. He recommends that Jewish immigrants who want to farm undergo a initial period of training that will teach them to acquire a love of the soil and an understanding of the need for working with neighbors in a community. Kirsch urges Jewish agricultural agencies to establish farm labor bureaus in cooperation with Federal and State governments to train the prospective settlers through a period of tenancy and agricultural education in order to reproduce the actual conditions they will have on a farm. He concludes by asserting that although the plan he presents may seem rigid, it is necessary because it gives the individual an opportunity to acquire the kind of training necessary to adapt to farm life with a greater chance of success.

382. Levine, Herman Joseph, and Benjamin Miller (1966). *The American Jewish Farmer in Changing Times.* New York: Jewish Agricultural Society. 100 pp.

A two part report on the early history of the Jewish farm movement and descriptions of the lives, work, and community involvements of the farmers and their families. Part I — The Past — Traces the history and development of Jewish agricultural settlement in America from the founding of the Baron De Hirsch fund in 1891 and the Jewish Agricultural Society in 1900 through the 1960s. The authors explore the changes and developments that transformed agriculture following WWII through the 1950's. Part II — The Present — reviews the conditions of Jewish farmers in the 1960's and the successful organization, marketing, production, and economic successes in New England, New Jersey, New York, the Midwest, the South, and California. The report concludes with facts and statistics regarding the role of the Jewish Agricultural Society, Jewish agricultural scientists in the United States, Jewish farm families, scholarship programs, and cooperative projects with Israeli agriculturists. The appendix provides an alphabetic list of directors and staff of JAS from 1900 through 1966.

383. Menken, J. Stanwood (1882). *Report on Formation of the First Russian Jewish Colony in the United States at Catahoula Parish, Louisiana.* New York: M. Thalmessinger (also London: Stephens, Hayter, and Co.). 18 pp.

A four part report presented to the Executive committee of the Hebrew Emigrant Aid Society (HEAS) of the United States describing the establishment of an agricultural colony for Russian immigrants on Sicily Island, Catahoula Parish, Louisiana. The colony is 350 miles from New Orleans and consists of one hundred and fifty recent immigrants. Menken begins with a description of the land, agriculture, and climate of the region, and an explanation of why that site was chosen. He gives details regarding terms of purchase, names and amounts of contributions from major benefactors who supported the settlement idea and donations of provisions, clothing, furniture, seeds, agricultural implements and other items needed to keep the colony functioning. The report also contains the proposed "Constitution for the First Agricultural Colony of Russian Israelites of America" with seventeen articles outlining the governance structure and members' rights and "What the Colony Consists Of" which offers brief information about the gender, ages and avocations of the settlers. It includes resolutions adopted by the citizens of Catahoula Parish welcoming the Jewish settlers and a letter from Herman Rosenthal, the President of the colony, thanking the New Orleans office of the HEAS for its assistance in establishing the colony. Menken concludes with his views on the importance of the settlement as a livelihood for the immigrants. (For a summary of Menken's report see: The first American Jewish colony. Interesting report of President Menken [1882 January 13]. *American Hebrew*, 9[9]:99–100.)

384. National Appeals Information Service (1929). *Jewish Agricultural Society.* (Reports on Jewish Institutions No. 9.) New York: Bureau of Jewish Social Research. 25 pp.

A descriptive and informational report prepared for the National Appeals Information Service covering the activities and expenses of the Jewish Agricultural Society. The data obtained from Gabriel Davidson, General Manager of the Society, and annual reports presents details about the purposes and administrative structure of the Jewish Agricultural Society with a list of its directors and managers; an outline of activities and costs of extension work, settlement, and farm loans, sources of income and receipts and expenditures.

385. Schwarz, Julius (1882). *Report of Mr. Julius Schwarz on the Colony of Russian Refugees at Cotopaxi, Colorado Established by the Hebrew Emigrant Aid Society of the United States, 1882.* New York: Hebrew Emigrant Aid Society of the United States. 18 pp.

A report from Julius Schwarz, General Manager of the Cotopaxi Colony (Colorado), to the President of the Hebrew Emigrant Aid Society. The report covers history of the colony, the topography of the region, statistics of the colony (number of families, ages, etc.), soil and land conditions, working capacity of the colonists, education and religious life, expenditures, and general remarks and conclusions. Schwarz concludes that the colony is an example of the perseverance of the settlers and a indication that Jews can be successful in agriculture.

386. Stainsby, William (1901). *The Jewish Colonies of South Jersey: Historical Sketch of Their Establishment and Growth.* Camden, NJ: Chew & Sons. 28 pp.

This illustrated report issued by the New Jersey Bureau of Statistics gives details about the successful agricultural colonies at Alliance, Rosenhayn, Woodbine, and Carmel and the Baron De Hirsch Agricultural and Industrial School at Woodbine which were supported by a variety of benevolent and philanthropic societies. The report describes how the land was acquired, number of settlers and allotment of tracts, soil conditions, crops, livestock, manufacturing, expenses and development. The Baron De Hirsch Agricultural and Industrial School was established in 1894 on a tract of land in the Woodbine settlement. It had an organized course of instruction that combined theoretical teaching and practical application. Details about enrollment, number and kind of buildings, curriculum, faculty, subjects taught, agricultural projects, and crops and livestock raised at the school are presented.

387. Zepin, George (1900). *Jewish Religious Education in Farm Districts.* Cincinnati: Union of American Hebrew Congregations. 13 pp.

A report to the Board of Managers of Synagogue and School Extension of the Union of American Hebrew Congregations regarding the lack of religious education for Jewish children on farms. Following a visit to some of the agricultural settlements, Zepin found that distances, lack of funds, and long work days have led to a neglect of religious education for farm children. He recommends that the Federation of Jewish Farmers and the Jewish Agricultural and Industrial Aid Society open a bureau to train teachers and *schochetim* (people skilled in methods of ritual slaughter and meat preparation) and offer subsidies to the settlements to hire and retain them. In Zepin's view the societies devoted all their efforts to making farmers out of Jews but overlooked the need for farm children to learn about their religious heritage.

Museum and Exhibit Catalogs

388. B'nai B'rith Jewish Historical Committee (1966). *Baron De Hirsch Fund 75th Anniversary Exhibit: April 15 to December 31, 1966.* Washington, DC: B'nai B'rith Jewish Historical Committee. 15 pp.

An information pamphlet accompanying an exhibit about the history of the Baron de Hirsch Fund held at the B'nai Brith Building in Washington, DC. The pamphlet gives a short biography of Baron Hirsch and his early life and business acumen. Describes in brief the founding of the Baron De Hirsch Fund, creation of the Industrial Removal Office, and the establishment of the Baron de Hirsch Trade School. Emphasizes Hirsch's agricultural colonization and education initiatives and its impact in acclimating Eastern European immigrants to life in America.

389. B'nai B'rith Klutznik National Jewish Museum (1992). *Agri-Culture: Jewish Immigrants, Farmers, Artists in New Jersey.* Washington, DC: B'nai B'rith Klutznick National Jewish Museum. 24 pp.

An illustrated catalog of an exhibition held at the B'nai B'rith Klutznick Museum in August 1992 celebrating the artistic legacy of New Jersey Jewish farmers. The catalog contains a short history of Jewish immigration to America and the farm families of New Jersey. The art was created by second and third generation descendants of the farm families. The exhibit contained photographs, paintings, drawings, prints, architectural drawings, illustrated books, calligraphy, stoneware, millinery, sculpture, woodcuts, ceramics, and jewelry.

390. Mintz, Matiyhahu, and Eliahu Benjamini (1983). *The Jewish Agricultural Experience in the Diaspora.* Tel Aviv: Beth Hatefutsoth—The Nahum Goldmann Museum of the Diaspora. 95 pp.

A brochure in Hebrew and English from an exhibition of 350 photographs, documents, and publications documenting Jews in agriculture around the world in the 19th and 20th centuries. Contains selected photographs and accompanying text arranged in geographical and chronological order showing Jews actively engaged in farming. Includes two photographs depicting scenes from the Woodbine, New Jersey settlement in 1894 and 1900.

391. Peck, Abraham J. (1986). *The American Jewish Farmer: An Exhibit Sponsored by the American Jewish Archives.* Cincinnati, OH: American Jewish Archives. 18 pp.

Catalog from an exhibit held at the American Jewish Archives in Cincinnati from May to November 1986. The exhibit contained over 150 photographs and documents depicting experiences of Jews in American agriculture from the beginning of the 19th Century focused on the history of the American Jewish farm experience and its evolution and eventual decline. The catalog includes eleven illustrations and three essays: "Jewish Agriculture in America" by Gertrude Dubrovsky which describes some of the early Jewish farm communities and their successes; "Jewish Agrarian Utopias in America: The Craving for a Great

Ideal" by Uri D. Herscher—an excerpt from Herscher's book *Jewish Agricultural Utopias in America, 1880–1910*; "A History of the Communitarian Settlement Known as 'New Odessa'" by Herman Rosenthal (translated from the Hebrew by Gary P. Zola), describing the New Odessa settlement in Oregon which existed from 1883–1886.

392. Ringwald, Christopher D., Harvey Strum and Jim Richard Wilson (1998). *Jewish Farming Communities of Northeastern New York.* Albany, NY: Rathbone Gallery, Sage Colleges. 48 pp.

This catalog prepared for the exhibit at the Rathbone Gallery (now Opalka Gallery) Sage Colleges, Albany, New York—February 23–March 22, 1998 contains five essays documenting the history and lives of Jewish farmers in Northeastern New York. "Jewish Farmers" by Harvey Strum (pp. 5–17) sets the stage with a chronological history of Jewish agricultural communities in the United States, Canada, Argentina, and Palestine between 1881 and 1914. "Fred Rheingold" by Christopher D. Ringwald (pp. 18–21) is a biography of a third generation farmer in East Shodack (Rensselaer County), who is also a ritual director and Hebrew Teacher at the Nassau Jewish Community Center. "Jewish Farmers of Northeastern New York" by Harvey Strum (pp. 22–24) offers a short history of Jewish farmers in Rensselaer county and their cooperative efforts in establishing credit unions, a fire insurance company, a farmers association. three synagogues, and a Jewish community house. "Oldtimers" (pp. 25–30) and "Newcomers" (pp. 31–33) by Christopher Ringwald provide glimpses into the lives of established farm families and some of the newcomers who have weekend homes or came to retire in the area. Includes acknowledgments, a bibliography, guide to the exhibit, and description of the Rathbone Gallery.

Film and Video

393. Burt, Bonnie, and Judy Montell (2002). *A Home on the Range: The Jewish Chicken Farmers of Petaluma.* VHS (52 minutes). Berkeley, CA: Burt/Montell Productions.

This is the story of a community of immigrants from Eastern Europe who came to America in the early 20th Century with socialist ideals and a progressive agenda. They wanted more out of life than crowded urban life and the sweatshops of New York. When news of a cooperative farming community in Petaluma, California reached the immigrants, many decided to join the community and become poultry farmers. They weathered conflicts and internal differences, community opposition, vigilantes, and House Un-American Activities Committee investigations and even criticism from other Jews to make their community thrive and become a major supplier of eggs to the United States markets. Interviews with former members, newspaper articles, and archival resources have been interwoven to portray a classic study of the challenges facing a community of Jewish immigrants who lived an worked together, participated in community affairs, and developed the agricultural industry in their community.

394. Davis, Charles (1998). *Kosher Valley.* VHS (60 min.). Boulder, Colorado: Throughline Productions.

A group of southern Colorado Mormon and Spanish ranchers facing bankruptcy decide to establish a kosher cooperative to sell their beef to a specialty market in hopes of increasing sales and raising the value of their cattle. They contact Rabbi Mayer Kurcfeld, who supervises the Star K kosher certification in Baltimore to educate them on the procedures for preparing and slaughtering cattle for the kosher market. Although problems and community opposition arises, the Rabbi gains their confidence and the co-op is launched. The ranchers establish a business model, contact potential grocers and butchers in Baltimore, and began running the new business. This video is a unique portrayal of how Americans of different cultures can work together for the mutual benefit of both communities.

395. Dubrovsky, Gertrude W. (1993). *The Land Was Theirs: Jewish Farmers and Their Communities.* VHS (58 Min). Waltham, MA: National Center for Jewish Film.

A nostalgic documentary about Jewish immigrants from Russia and Germany who settled on farms in New Jersey during the early 20th century raising chickens and eggs. Focuses mainly on Farmingdale, Perrineville, Freehold, Toms River, Hightstown, and New Brunswick areas. Surviving members of the farms, give oral histories about their farms and the secular, religious and cultural experiences that shaped their lives. Includes interviews with Jewish families that are still farming in New Jersey and plan to keep the farms in the family as a legacy to their offspring. Dubrovsky, who was raised on a chicken farm in Farmingdale, NJ, which was condemned in 1970 to construct a reservoir, provides narrative that gives viewers a historical understanding of the Jewish farmers and their physical, cultural, and social settings.

396. Gerber, Scott, Bonnie Burt and Judith Montell (2002). *Song of a Jewish Cowboy.* DVD (18 Minutes). Berkeley, CA: Burt/Montell Productions.

This is a biography of Scott Gerber who was raised in the community of chicken farmers in Petaluma California. He still retains the left wing ideals of the community and the songs relating to the various causes the community participated in which he learned from his mother and grandmother. He sings cowboy and Yiddish songs and talks about how he copes with his life as a cowboy and bridges two different worlds.

Miscellaneous — Speeches, Catalogs, Archives, and Special Publications

397. American Hebrew Agricultural Association (1855). *Constitution of the American Hebrew Agricultural Association.* New York: J. Murlhauser. 11 pp.

Contains the aims and objectives of the Association adopted October 5, 1855. A brief introduction indicates that the founders hope this will encourage Jews to enter the noble profession of agriculture. It then presents the structure and governance of the organization in the form of a constitution. No further information is given regarding goals or activities.

398. Bailey, Christine Howard (1986). *Mrs. Mehl and the Chicken Hawk: An Oral History Presentation Concerning the Colchester Jewish Experience, 1880–1930.* 27 pp.

A group of Russian immigrants came to Colchester, Connecticut in 1880 and became farmers. They remained in the community and continued to make their livelihood in farming despite economic hardships and conflicts regarding their religious, cultural, and social traditions. Over time they became an important part of the Colchester community and were instrumental in development of the area. This playbill represents a play based upon oral histories of the Jewish immigrants and two generations of descendants. The play is compiled from secondary texts, local land and school records, and oral history interviews with individuals from the Colchester community in the collection of the Greater Hartford Jewish Historical Society. The playbill contains information about the cast and crew, a history of the Colchester Jews, and how the play was created and presented.

399. Baron de Hirsch Agricultural and Industrial School (1898). *Catalogue of the Baron De Hirsch Agricultural School, Woodbine, New Jersey.* New York: De Leeuw & Oppenheimer Printing Co., 24 pp.

A catalogue for prospective students describing the departments, faculty, courses, admission requirements, school library, school farm, map of the grounds, and regulation for retention in the school.

400. Baron de Hirsch Agricultural and Industrial School. (Dec. 1900–Jan. 1906) *Hirsch School Journal.* Woodbine, NJ: Baron De Hirsch Agricultural and Industrial School.

An informational journal about the Agricultural and Industrial School. Contains directory, editorials, school notes, alumni notes, student life, the school farm, and other features.

401. Baron de Hirsch Fund (1906). *Baron De Hirsch Fund: Trust Deed, Certificate of Incorporation, By-Laws and Trustees.* New York: Oppenheimer Printing Company. 28 pp. (Revised and reissued New York, Baron de Hirsch Trade School Press, 1924.)

This four part document contains the deed of trust explaining the use of Baron de Hirsch Society funds and how to distribute them, the objectives and purposes of the fund, number of trustees and seventeen by-laws regarding memberships, duties, and administration of the fund.

402. Cowen, Philip (189? not dated). *Colony Alliançe: Its Founding and Progress. Can the Jew Be an Agriculturist?* New York: Press of Philip Cowen. 9 pp.

A pamphlet by the founder and editor of *American Hebrew* distributed at "Palestine Bazaar," an exhibit and sale of the agricultural products produced at the Alliance Colony. Presents a brief history of the founding of Alliance, sponsorship and funding, numbers and characteristics of settlers, progress and expansion, explanation of the change from cooperative settlement to individual ownership, and a list of the produce for sale at the exhibit.

403. Davidson, Gabriel (1940). *Who Is the Jewish Farmer? A Sociological Study.* New York: Jewish Agricultural Society. 4 pp.

Summary of a survey among Jewish farmers to determine who they are, where they came from, are they successful, and are their families contented with farm life. The Society distributed questionnaires to 1225 Jewish farm families around the country and received 395 replies (a 32 percent response rate). The study revealed a wealth of information regarding their backgrounds, experience, attitudes about farming, relations with gentile neighbors, and involvement in Jewish life. The overall responses were favorable with respondents indicating that despite economic problems and hard work, they were satisfied with their lives on the farm and many hoped their children would remain in agriculture.

404. Dubrovsky, Gertrude W. (1976). *The Farmingdale Collection.* New York: YIVO Institute for Jewish Research. 12 pp.

A guide to archival materials in the YIVO Collection covering the Jewish farming community of Farmingdale, NJ from 1919 through 1960. The collection includes taped interviews in English and Yiddish with 116 Jewish farmers, private papers, letters, journals and photographs, video and audio tapes, and paintings by settlers and their families, maps, mortgages, deeds, tax maps, and other official records, of the community's cultural, social, economic, educational, and fraternal organizations, the Jewish community center and newspaper accounts of community activities in the local and Yiddish newspapers. (See also: Perrone, Fernanda H. *A Guide to the the* [sic] *Farmingdale Collection, 1890–1952.* Rutgers University Libraries; Special Collections and University Archives. 1976.)

405. Federation of Jewish Farmers of America (1916). *Farm Almanac and Buyers' Guide.* New York, N.Y.: The Federation of Jewish Farmers of America. 130 pp.

This almanac published mainly in Yiddish contains over 40 articles on all aspects of farming and farm life and related information. The agricultural articles cover topics of interest to farmers including feeding calves, grain and field crops, spraying for insect protection, apple production, veterinary medicine, etc. Related topics deal with the Federation of Jewish Farmers, farmers and boarders, immigration, farm labor, savings and loan associations and other topics of importance. Articles in English discuss the Federation and list of branches, the Jewish Agricultural and Industrial Aid Society, and rural credit.

406. International Society for the Colonization of Russian Jews (1891). *The Occupation of Agriculture as a Means for the Betterment of the Condition of Russian Jews.* San Francisco: International Society for the Colonization of Russian Jews. 11 pp.

A paper announcing the establishment of the International Society for the Colonization of Russian Jews and solicitation for the sale of shares to help carry out the mission. The goal of the Society is to establish an agricultural colony for Russian Jews in the "semi-tropics" which will allow them to be independent and industrious. The paper points out that Jews have been involved in agriculture throughout history but restrictions on land ownership posed hardships that caused them to seek other professions. The Society believes that the Russian Jews should pursue agriculture instead of trade or barter as the road to material and spiritual well being. Gives 12 reasons for choosing the semi-tropics as the best location for the settlement.

407. Jackson, S. H. (1837). *Address and Articles of the Association Zeire Hazon to Their Brethren in the United States.* New York: The Association. 3 pp.

The association Zeire Hazon (Tender Sheep) was attempting to gain support for organizing an agricultural settlement on the Western prairie for German Jewish immigrants. This speech presented to the Association at their meeting in Spring, 1937, explains the reasons for the plan and how it will be implemented. (Also published with commentary as: "Agricultural Project. Address to the Jews of the Association Zeire Hazon [Tender Sheep], New York, Spring 1837" in: Schappes, Morris U, editor. *A Documentary History of the Jews in the United States, 1654–1875.* New York, Shocken Books, 1950 and 1971. Pp. 195– 198.)

408. Jewish Poultry Farmers Association of New Jersey (1962). *Jewish Poultry Farmers Association of New Jersey. 10th Anniversary Journal.* New York: Futuro Press. 48 pp.

A journal celebrating the 10th anniversary of the Jewish Poultry Farmers Association. Published in English and Yiddish, contains articles on the Organization's development and its services to the poultry farmers, historical accounts and reminiscences from the Jewish farmers of Vineland, and greetings from organizations, individuals, and businesses that serve the community.

409. Krauskopf, Joseph (1918). *Hunger as Great a Menace as the Hun. Address at the National Farm School by Its Founder and President, the Rev. Dr. Joseph Krauskopf, September 29, 1918.* Doylestown, PA, National Farm School. 32 pp.

This speech was given to an audience of potential donors seeking monetary support for the National Farm School. Krauskopf focuses on the nation's food needs during World War I and the necessity of averting a food scarcity following the war. He emphasizes the contribution of the National Farm School as a training institution that will train able bodied men between the ages of eighteen and forty five for agricultural occupations. He expresses his wish to establish housing and facilities for girl students. He provides an update about the school and its programs and facilities including a poultry plant, botanical laboratory, and a new farm. Krauskopf points out the school has gained strength and vitality and reminds the audience that their financial support will keep the school going. Includes a list of life members and contributors for the current year.

410. Krauskopf, Joseph (1915 September 26). *After the War—What?* Doylestown, PA, National Farm School. 12 pp.

In this speech at the National Farm School, Krauskopf points out that agriculture is the vocation of peace. The war signals the need to demonstrate to the Nation's youth the importance of agriculture and the eagerness to respond to "the call of the soil." In this appeal to supporters for funds to enable the National Farm School to train urban youth for agricultural work, Krauskopf describes the work of the school, the achievements of successful graduates, his plans for adding new departments and facilities, and admitting girls. He urges supporters, friends and Jewish organizations to come forward and relieve the school of its debts so that it can continue to perform its vital role in training Jewish youth for a life in agriculture.

411. Kurtz, Birdie (1979). *A Capsule History of the American Jewish Experience in Toms River, New Jersey in the First Half of the Twentieth Century: Citizen, Farmer, Jew, Zionist.* Matawan, NJ: B. Kurtz. 18 pp.

In the late 1930s, Benjamin and Birdie Kurtz, bought a 19-acre alfalfa field on Dugan's Lane in Toms River, New Jersey with assistance and support from the Jewish Agricultural Society and began raising chickens. Over the years, their farm grew to a successful enterprise with 5000 chickens. Kurtz presents a nostalgic look back at life on the farm and the community. She discusses care of the chickens, relationships with other Jewish and Christian farmers, egg marketing, raising children, the Jewish Community House and Jewish education, the secular high school, support of Jewish organizations, and other community services and events.

412. Lee, Francis Bazley (1893). *Address on New Agricultural Improvement in Southern New Jersey. Delivered by Francis B. Lee, Before the State Board of Agriculture.* Trenton, NJ: The John L. Murphy Publishing Company. 8 pp.

Lee, a New Jersey attorney and historian, presented a speech before the State Board of Agriculture offering his observations on the Jewish agricultural settlements in southern New Jersey. Lee focused on Woodbine and referred to the settlers as "Hebrews." His speech covered the structure of Woodbine with descriptions of the houses, roads, and farmlands and the suitability for settlement by Hebrew farmers. He concluded with opinions about the "personal qualities" and political and religious practices of the settlers and their willingness to adapt to the way of life in the communities where they live.

413. Menkes, Jacob Bernard (1933). *Judah's Sons of the Soil: An Historical Summary Dealing with the Jews in Farming. Accompanied by Bibliography and Questions for Discussion.* Cincinnati, OH: National Federation of Temple Brotherhoods. 19 pp.

A comprehensive analysis of the involvement of Jews in agriculture from biblical times to the present intended as a study topic for synagogue youth and adults. Menkes takes readers through the ages beginning with references to agriculture in the Bible and Talmud. He points out that Jews were farmers in Assyria, Persia, Asia Minor, Egypt, and Rome. Jews were engaged in farming in France, Spain, Poland and other European countries through the end of the 18th century. In the 19th century, Russia and Turkey had large numbers of Jews on the farm and early settlers in Palestine considered agriculture as a foundation of nationalist and Zionist progress. Then focusing on the "New World," Menkens points out that Jews were involved in agricultural pursuits during colonial days and this momentum continued through the 18th and 19th centuries with attempts to establish colonies and independent farming communities in the United States, Canada and South America. He concludes with ten review questions asking about the prospects for the future of Jewish farming and how it is shaped by historical events.

414. Miller, Mrs. Ben, and Daniel S. Roher (1959). *50 Golden Years: The Ellenville Hebrew Aid Society, 1907–1957.* Ellenville, NY: Ellenville Hebrew Aid Society. 116 pp. (Published in observance of the Fiftieth Anniversary of the founding of the Ellenville Hebrew Aid Society, by the Board of Directors.)

The Ellenville Hebrew Aid Society provided social, economic, and religious assistance to Jewish immigrants who began to settle in Ellenville, New York in the early 1800s. This report presents a nostalgic history of the Society and the Jewish community of Ellenville. Several articles present the history of Ellenville's Jewish farmers and their participation in community affairs, town government, and religious life. These include "Ellenville's Early Jewish Families" (p: 27–31); "The Hebrew Aid Synagogue" (pp. 47–51); "The Jewish Farmers' Contribution to Ellenville's Growth" by Herman J. Levine, Former Manager, Ellenville Office, Jewish Agricultural Society (pp. 52–53); and "Leurenkill farmers start a Hebrew School" by Ephraim Yaffe (pp. 59–60).

415. Noah, Mordecai M. (1905). *Mordecai M. Noah's Discourse on the Restoration of the Jews. Republished in Extract with an Introductory Note by D.S. Blondheim.* Lord Baltimore Press: Baltimore. 8 pp.

Mordechai Manuel Noah (1785–1851) was born in Philadelphia of Portuguese Jewish descent. He was a trader, lawyer, and politician who traveled heavily in Europe and observed the contrast between the position of Jews in Germany, Poland, Russia, and the United States. Noah recommended establishing Jewish settlements in Syria, the territory surrounding Jerusalem, and the United States. He was well known for his plan to establish a settlement at Grand Island, New

York near Buffalo to be called "Ararat." This is an excerpt of his discourse given at an Episcopal Church in Buffalo, NY on September 2, 1825, reprinted from *The Maccabaean,* April, 1905. (Originally published as: *Discourse of the Restoration of the Jews: Delivered at the Tabernacle Oct. 29 and Dec. 9, 1844.* By M.M. Noah. With a map of the land of Israel. New York, Harper & Brothers, 1845. Also published with commentary as "Ararat, City of Refuge, 1825–1826," in: *The Jews of the United States, 1790–1840: A Documentary History,* Blau, Joseph L. and Salo W. Baron, editors. New York, Columbia University Press, 1963. Vol. 3, pp. 894–905.)

416. Peixotto, Benjamin F. (1887). *What Shall We Do with Our Immigrants?* New York: Young Men's Hebrew Association. 24 pp.

In this address to the Young Men's Hebrew Association of New York on February 2, 1887 commemorating the birth of Sir Moses Montefiore, Peixotto points out that large numbers of "Israelites" are migrating to the United States from Roumania and Russia. He discusses the discrimination they encountered in those countries and urges the audience to welcome them. He asserts that the answer to the question of how to absorb the Jewish immigrants is to encourage them to settle in agricultural colonies. Peixotto gives examples about successful colonies that have been established in various states. Appendix I, "Jewish Colonies in America" (pp. 17–20), outlines some of the features of colonies in New Jersey, Kansas, Virginia, and Oregon. Appendix II, "How to Promote Agriculture Among the Jews" (pp. 21–24), explains how to establish colonies and promote the concept to the new immigrants.

417. Perrone, Fernanda H. (1996). *A Guide to the the [sic] Farmingdale Collection, 1890–1952,* 32 pp.

A guide to the materials in Special Collections and Archives at Rutgers University Libraries covering the history of the Jewish farmers in Farmingdale, New Jersey during the period 1890–1995. Most of the materials cover 1919 to 1976 and were collected by Gertrude Dubrovsky for research on her book *The Land Was Theirs: Jewish Farmers in the Garden State* (University of Alabama Press, 1992). The collection includes biographical surveys, oral history transcripts and audiotapes, photographs, newspaper clippings and personal mementos and documents. (See also: Dubrovsky, Gertrude W. *The Farmingdale Collection.* New York, YIVO Institute for Jewish Research, 1976.)

418. Petaluma Jewish Community Center (1975). *Petaluma Jewish Community Center 50th Anniversary Souvenir Journal.* 44 pp.

This illustrated souvenir journal commemorates the 50th anniversary of the Petaluma Jewish Center established in 1925, Contains greetings from the Rabbi and officers of the Center, brief histories of the activities, religious school, clubs, chorus, B'nai Brith Women, Pioneer Women, Hadassah and other activities. Includes reprints from the tercentenary program covering the history of the first 100 years of the community, some of the farm families who were among the early settlers, and details about the Haas Memorial Loan Fund that provided loans to those interested in farming in Petaluma.

419. Petaluma Jewish Community Center (1954). *Tercentenary Souvenir Program Commemorating 300 Years Jewish Settlement in America.* Petaluma, CA: Petaluma Jewish Community Center. 62 pp.

This souvenir program was issued by the Petaluma Jewish Community during the commemoration of 300 years of Jewish settlement in America. The program contains a series of articles regarding the community and a wide spectrum of cultural and religious activities serving the Jews of Petaluma. Includes a history of Petaluma from 1864 through 1954 and the prominent role of Jewish farmers in the early years of the community.

420. Provisional Commission for the Establishment of Jewish Agricultural Settlements (1933). *We Jews Must Reconstruct Our Lives.* 5 pp.

The Provisional Commission was formed during the economic depression in response to the back to the land movement among small groups of Jews who wanted to live and work on the land. This pamphlet issued by a joint effort of the Jewish National Workers Alliance, Workmen's Circle, and United Hebrew Trades appeals to national organizations and civic leaders to provide financial assistance to develop Jewish agricultural cooperative colonies in accordance with modern agricultural perspectives. The Commission recommends the establishment of an Agricultural Finance Corporation to develop and oversee the colonies. Outlines a tentative plan of administration and finance of modern settlements and states that a well designed plan will help those who want to pursue a livelihood in agriculture to accomplish that goal.

421. Robinson, William Davis (1819). *Memoir Addressed to Persons of the Jewish Religion in Europe on the Subject of Emigration to, and Settlement in, One of the Most Eligible Parts of the United States of North America.* London: H. Hay. 40 pp.

Robinson, a Christian, was concerned about the persecution and religious intolerance directed against Jews in several parts of Europe and Turkey. He issued this memoir urging wealthy European Jews to purchase land in the upper Mississippi and Missouri territories to make it possible for Jews to emigrate and pursue a life in agriculture. The land would be offered on credit along with transportation and funds to buy basic tools and equipment. Robinson outlines his six step plan to achieve this objective and suggests that this will provide Jewish immigrants a life in the United States in an atmosphere of "a rich soil, a smiling country and a congenial climate." (For an excerpt of Robinson's plan with commentary, see: *"Invitation to Immigration"* in: Schappes, Morris U., editor. *A Documentary History of the Jews in the United States, 1654–1875.* New York, Shocken Books, 1950. Pp. 141–147, and "Mr. W.D. Robinson's Helping Hand" in: *The Jews of the United States, 1790–1840: A Documentary History,* Blau, Joseph L. and Salo W. Baron, editors. New York, Columbia University Press, 1963. Vol. 3, pp. 879–884.)

422. Sampson, Michelle R., and Korelitz, Seth (1870–1991). *Baron De Hirsch Fund. Records. 1870–1991.* New York: American Jewish Historical Society.

Records of the fund's activities located in 196 manuscript boxes at the American Jewish Historical Society in New York. The Hirsch fund was devoted to help Jewish immigrants after they arrived in the United States. The Fund's philanthropic efforts were focused on agriculture, trade schools, and subsidies and grants to several Jewish agencies. The agricultural philanthropy covered the Woodbine Colony (New Jersey) 1891–1944, the Woodbine Agricultural School (1893–1917), and the Jewish Agricultural Society and the Jewish farming colonies (1900–1972). The boxes contain biographies, historical records, administrative reports, correspondence, loans, mortgages, annual reports, investment records, meeting minutes, maps, blueprints, photographs and other personal and public documents regarding Baron De Hirsch and the Fund's activities. (Guide to contents of the records: http://www.cjh.org/academic/findingaids/AJHS/pdf/Baron_de_Hirsch_Fund.pdf.)

Title Index

References are to entry numbers.

Subject Index

References are to entry numbers.